PROTOZOA
A Poseidon Adventure!

by Ellen J. McHenry

Published by Ellen McHenry's Basement Workshop in State College, Pennsylvania, USA.
www.ellenjmchenry.com
ejm.basementworkshop@gmail.com
Printed by Lightning Source, Ltd., La Vergne, Tennessee, USA.

ISBN 978-0-9887808-5-9

TABLE OF CONTENTS

INTRODUCTION

Pusillus Poseidon Adventures... "Directions, Rates & Fees, Latin & Greek, Contact..." — Latin and Greek?!

Yes, there's a button that says "Latin & Greek." — Click on it.

"Pusillus" is Latin for very small. Poseidon was the Greek god of the sea. — What in the world??

Want to go see what it is? — Sure! Click on "Directions."

S E V E R A L H O U R S L A T E R ...

Are you sure this is the correct address? — Look- the sign says, "Pusillus Poseidon Adventures."

There's nothing here. What's going on? — I don't know. Let's go ask that guy...

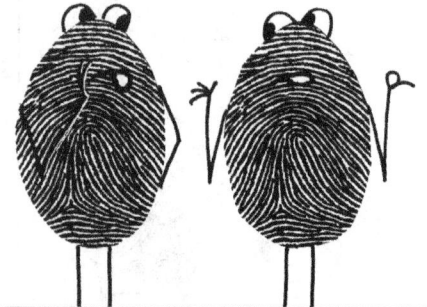

Um... we're confused. We thought this location offered "the adventure of a lifetime." — No mistake! This is it!!

I understand why you'd call this place "pusillus," but it doesn't exactly conjure up images from Greek myths! — Oh- you just wait and see!

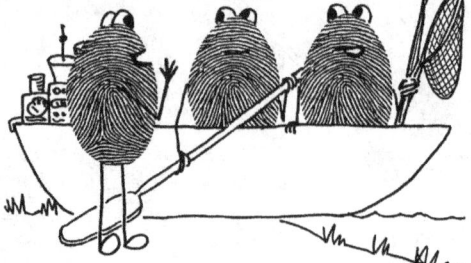

Press the green button to start the shrinking rays, and push the red button to stop them and return to normal.

Got it!

Here we go!

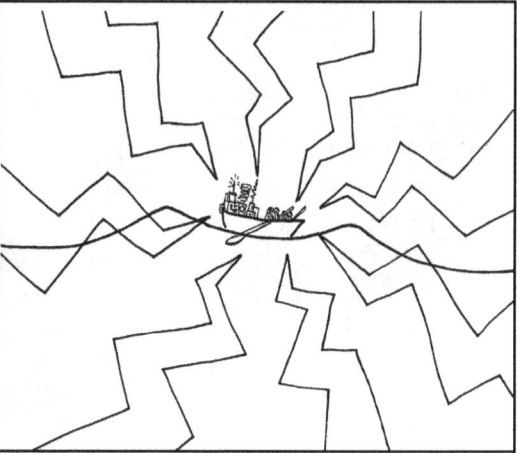

Yikes! Green icebergs? I don't remember seeing stuff like this in the pond!!

Remember, we're miniature now. Those things are probably duck weed leaves. I do remember seeing duck weed floating by the edge of the pond.

How small are we?

Um... I think if you look up, it will put things into perspective.

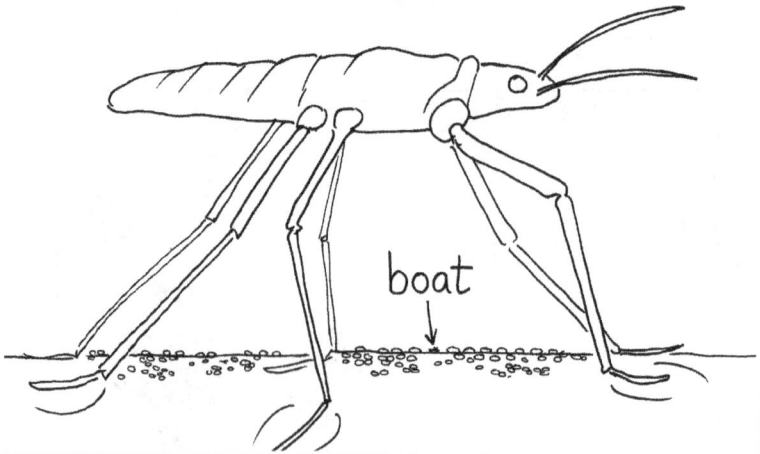

boat

Is that one of those teeny tiny water bugs? Because if it is....

...yeah, that would mean we're REALLY small!

Do you think we're, like, microscopic?

I don't know. Maybe we'd better start reading?

CHAPTER ONE: MICRONS AND PARAMECIA

How do we measure really tiny things? For example, how would you measure the dot underneath the question mark at this end of this sentence? It's pretty small. If you put a ruler next to it, you'd find that centimeters or inches would be far too large. You'd have to estimate in fractions, making it very difficult to be accurate.

The smallest lines on this ruler are **millimeters**. (A millimeter is 1/1000 of a meter. A meter is about as tall as an adult's waist.) As you can see, there are ten millimeters in a centimeter. (In other words, a millimeter is a 1/10 of a centimeter.) A millimeter would be a better measuring unit to use than a centimeter, but let's be honest—even a millimeter is still pretty big in comparison to that little dot. Is the dot half a millimeter? A fourth? A tenth? It's hard to tell. You'd need a magnifier to get a good estimate. And even then, you'd be taking a guess.

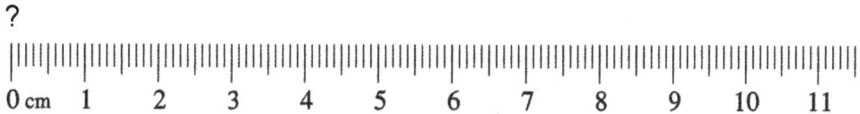

?

|||||||||||||||||||||||||||||||||||||||||||||||||||||||||||||||||||||||||||||||||||||||||||||||||||||
0 cm   1   2   3   4   5   6   7   8   9   10   11

Looks like we need a unit of measurement smaller than a millimeter. How small should it be, and what should we call it? We could divide a millimeter into even smaller units, like tenths or hundredths or thousandths. Dividing it into tenths might allow us to measure that dot fairly accurately, but what about even smaller things? What if we wanted to measure a bacteria?

Scientists decided to divide the millimeter into 1,000 smaller units called **micrometers**. This makes sense because you will need a <u>micro</u>scope to see things that are measured in <u>micro</u>meters. Now what about an abbreviation? "Millimeters" is abbreviated as **mm**: "m" for "milli" and "m" for "meter." If we used the same method for micrometers, we come up with... mm. Oops. We can't have two measurements with the same exact letters. We won't be able to keep them straight. What to do?

Zeus to the rescue! (Well, not really. Just the ancient Greeks.) The Greeks had basically the same alphabet as we do, but some of the letters looked just a little different. For example, their letter "m" in the lower case looked like this: μ (It might look a "u" but it is an "m" and they called it "mu.") If we use the Greek m, we can still use "mm" for micrometer, but it will look like this: μm. That way we won't get it mixed up with the mm that stands for millimeter. When we see "μm" we'll say "micrometer." Or, even better, we could use a shorter and easier word that means the same thing. How about **micron**?

Zeus was the king of all the mythological Greek gods and goddesses. He has no connection to the words micrometer or micron. But he did know Poseidon.

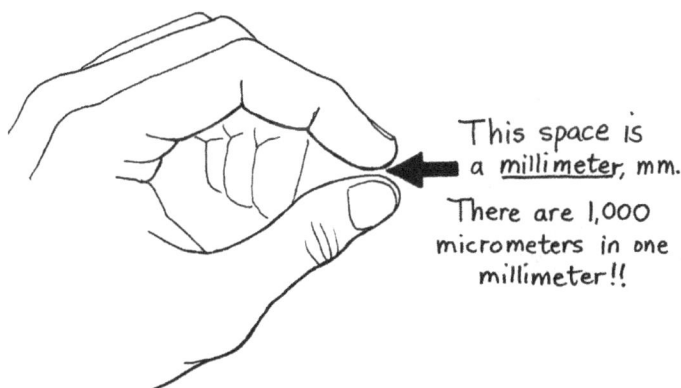

So, how big is a micrometer (a micron)? To get your mind around how small these units are, put your finger and thumb so that they are almost touching, but don't let them actually touch. That tiny space is a millimeter—one of those little increments on the ruler above. There are 1,000 microns in that little space. Many bacteria are only one micron in diameter. That means that you could fit a string of 1,000 bacteria into that space!

This space is a <u>millimeter</u>, mm.

There are 1,000 micrometers in one millimeter!!

So... how big—or small—are we? Are we in microns?

You, the reader, have a copy of their guide book, located after the last chapter of this book.

Take a quick look at the guidebook. You'll see that it has basic information arranged in a very structured format. The guidebook will come in handy when you need to find basic facts quickly. For example, if you are doing an activity and need to compare sizes or feeding behaviors, the guidebook will be much easier to use than this text. (If you have a hard copy of this book (instead of digital) and you would like to print additional copies of the guidebook, you can download and print by going to www.ellenjmchenry.com and clicking on FREE DOWNLOADS, then on MICROBIOLOGY, then on PROTOZOA GUIDEBOOK.)

The **Paramecium** was probably the very first **protozoan** to be discovered. The Dutch scientist Antony van Leeuwenhoek saw these in his simple microscope in the late 1600s. Then, in 1718, a French scientist named Louis Jablot (*Zhah-blo*) published a description of a little "animal" he had found in a drop of water. Jablot called this creature a "chausson" (*sho-SOHN*) meaning "slipper" because its shape does look a bit like a slipper. The name stuck, and this creature was referred to as the "slipper animacule" for the next two hundred years.

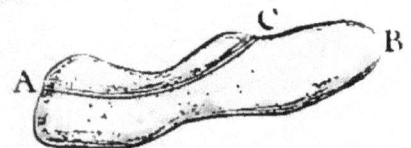

Jablot's drawing of the "little slipper"

The official name, "Paramecium," was created in 1752 by English scientist John Hill, using the Greek word "paramekes" meaning "oblong." Hill used this term to describe any microscopic creature that had no fins, legs, tails, or any other visible appendages. Few people actually used this term, however, and continued to use the very cute nickname "little slipper."

In 1773, a Danish scientist named Otto Müller changed the spelling to "Paramoecium." Then, in 1838, a German scientist decided to change the spelling back to "Paramecium," and that is what we still use today. In future chapters, we will see other places where scientists have disagreed about how to spell words. The fight about whether to use "oe" or just "e" will show up again soon. Strangely, in some cases, the "oe" is still with us, though in very recent times there has been an effort to standardize spelling and get rid of "oe" once and for all. But enough about that for right now... back to Paramecia. (Paramecia is the plural form of Paramecium. One Paramecium, two Paramecia.)

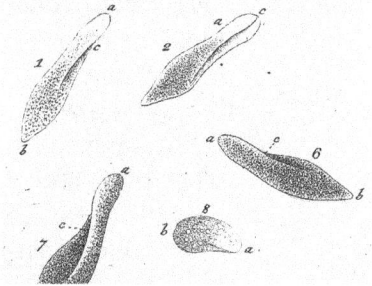

Muller's drawing of Paramecia

What are these strange creatures? Even though they look and act like little animals, they are not in the animal kingdom. To be a part of the animal kingdom, you have to be ***multicellular***, meaning made of many cells. The Paramecium is made of only one cell, so it is ***unicellular***. It doesn't have skin cells or blood cells or neurons or any of the specialized cells that animals do. It is just one cell. However, this one cell is very much alive and does many things you do every day: move, find food, eat, digest, take in oxgen, expel waste, move away from uncomfortable situations, defend itself, and even communicate with others of its species. How does it accomplish all of these things without any organs such as muscles, stomach, heart, lungs, brain or kidneys?

Let's take an up-close look at the anatomy of a Paramecium. One nice thing about studying protozoa is that they are transparent. You don't have to cut them open to see inside.

A Paramecium doesn't have skin, but it does have an outer layer called a ***pellicle***. The word pellicle comes from the Latin word "pellicula" meaning "skin" or "husk." If you've ever "husked" corn, you've peeled off what the Latin-speaking Romans would have called a pellicle: a protective layer designed to cover something important inside.

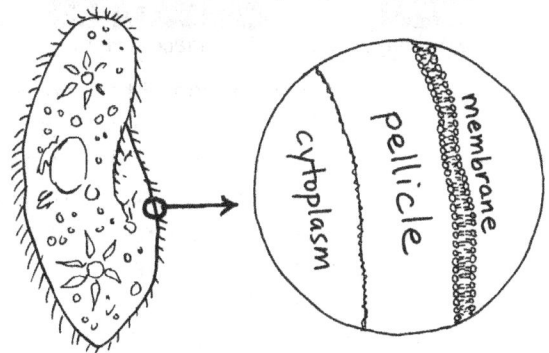

The pellicle is very thin (less than a micron) but has several layers. The inner layers are made of a substance that is tough but flexible. The body of the Paramecium needs to hold its shape, but it must also be able to bend and twist so it can get out of tight spots.

The outermost layer is an extremely thin ***membrane*** (only two molecules thick!) that acts like a fence, or screen, all around the cell. It keeps large molecules from entering the cell, but will allow tiny molecules, such as water and oxygen, to leak through. (Large molecules that are helpful can be brought in at places that act as gates.) This type of membrane (often called a ***plasma membrane***) is found not just in protozoa but in all living cells. Every cell in your body is surrounded by a plasma membrane.

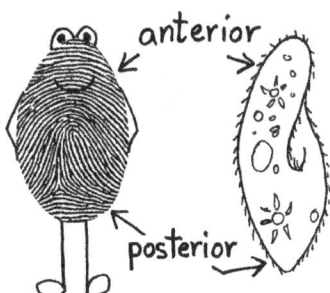

The Paramecium has openings in its pellicle, just like you have openings at various places in your body where there is a hole that leads to the interior: mouth, nose, ears, and... some "exits" at your posterior end. ***Posterior*** means your bottom part. ***Anterior*** means your top part. ("Post" is Latin for "after" and "ante" is Latin for "before.") Your head is located at the anterior region of your body. If you want a polite way to refer to your bottom, you can call it your posterior.

Paramecia also spiral as they swim forward.

The Paramecium's anterior isn't a head. It does not have eyes or ears or a brain or anything that you would recognize as part of a head. What makes scientists consider one of its ends as the anterior is that when it swims, a Paramecium will generally move in the direction of its anterior end, as if it were a fish or snake or something that moves headfirst.

Another similarity between your skin and a Paramecium's outer covering is that they both have hairs. The Paramecium's hairs are called *cilia*. This word comes from Latin and means "small hairs or eyelashes." (The singular is cilium. One cilium, two cilia. Just like Paramecium and Paramecia.) The Paramecium can move its cilia like little arms or fins, propelling itself through the water. The cilia are hard to see, though, unless you have a microscope with very high magnification. If you are lucky enough to have an extremely powerful (and extremely expensive) kind of microscope called a *scanning electron microscope (SEM)*, you can look at the surface texture and see all the cilia.

A scanning electron microscope (SEM) uses electrons to "see" instead of light. You don't look into an electron microscope; the chamber is all sealed up. The microscope bounces electrons off the sample, causing the electrons to go flying off at various angles. The electrons then hit a screen where they are recorded. The pattern of all these millions of hits on the screen eventually makes a picture that looks three-dimensional. The drawback to this method is that you can't see anything inside the Paramecium. You only see the surface texture. So it really takes both regular microscopes and electron microscopes to give us an accurate understanding of what a Paramecium is like. If you see just one or the other, you miss important visual information.

An SEM image showing the surface texture

The Paramecium moves its cilia using the same basic chemical process that your muscles use. Movement is caused by chemical changes (usually a sudden increase in calcium). This chemical change causes protein chains in the cilia to contract.

The motion of the cilia is surprisingly similar to how we use our arms to swim. There is a forward "power" stroke that propels the body, then a weak "reset" stroke where the cilia (or arms) are brought back to their original position. The Paramecium can swim backwards, too, if it needs to. If it bumps into something it can back up, turn, then try going forward in a new direction.

A Paramecium's swimming motion looks very smooth. The cilia don't all beat at once; they take turns in a very orderly fashion to create waves of motion. Beating cilia might look a bit like a gust of wind blowing over a field of tall grass. If you'd like to see Paramecia swimming, Activity 1.1 gives directions for accessing links to videos of Paramecia.

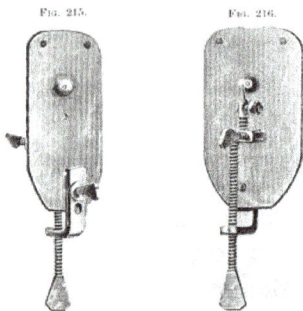

The world's first microscopes, made by Antony van Leeuwenhoek in the late 1600s.

This is the first microscope to use two lenses, at the top and bottom of the tube. It is from the 1700s.

A modern SEM "electron microscope," showing the chamber open. The chamber will be sealed and the air removed.

The Paramecium's pellicle can't feel things like your skin does. Your skin is loaded with nerve cells that sense hot, cold, pain and pressure. The Paramecium is just one cell so it can't have nerve cells. But even without nerve cells it is somehow able to sense things in its environment. If the environment around it becomes too hot, cold, acidic, or toxic, it will try to move to a place that is safer.

A Paramecium also somehow knows how to search for food, even without a brain to give it hunger signals or to tell it what to eat. It "knows" what it can digest: bacteria, algae, plant cells, and even other ciliated protozoa (smaller ones). The place where it takes in food is called the **oral groove**. ("Oral" comes from the Latin word root "or-," meaning "mouth.")

At the bottom of the oral groove is an opening which is sometimes referred to as its mouth. The food goes through this opening into a short tube called the **gullet**. The gullet is the rough equivalent to our throat (esophagus).

At the bottom of the gullet a bubble sort of thing starts to form, collecting the food. The more technical word for a bubble inside a cell is a **vacuole**. (You might notice the similarity of this word to the word "vacuum." Both words come from the Latin word "vacuus," meaning "empty.") Once this vacuole fills up, it pinches off from the gullet and goes floating away into the interior of the cell. While it circulates around the cell, digestive enzymes will enter the vacuole and break down the food. This is exactly what happens in your stomach. Your body makes chemicals that break down your food into tiny bits that your cells can absorb. So the food vacuoles are sort of like little stomachs.

The digested nutrients leak out of the vacuoles into the cellular fluid, called the **cytoplasm**. ("Cyto" comes from the Greek word "kytos" meaning "container," while "plasm" is from the Greek word "plasma" meaning "something molded or formed." The word plastic is also related to the word "plasma." Plastic is definitely able to be molded and formed into things.) The cytoplasm is a jelly-like substance made mostly of water. All cells, including every cell in your body, is filled with cytoplasm. Dissolved gases such as oxygen and carbon dioxide float around freely in the cytoplasm, as do small nutrients such as glucose (a sugar).

Notice that the Paramecium has an **exit pore** where it can get rid of waste. This is somewhat similar to the exit at the end of your digestive system, although not nearly as complicated. The Paramecium's exit is basically a vacuole that opens to the outside.

A Paramecium is constantly taking in food. As soon as one food vacuole fills up and pinches off from the gullet, a new one starts to form. A Paramecium may have many food vacuoles floating around inside. The cilia along the oral groove sweep food down toward the gullet. Along with the food, water gets swept in, too. This constant sweeping motion brings lots of extra water into the cell. The Paramecium needs a way to get rid of the extra water. It does not have kidneys or a bladder, but it does have two **contractile vacuoles**. They are easy to spot because they are star-shaped, with arms radiating out from a central circle. The arms gather the extra water and bring it to that central circle, which then acts like a pump, pushing the water back outside of the cell.

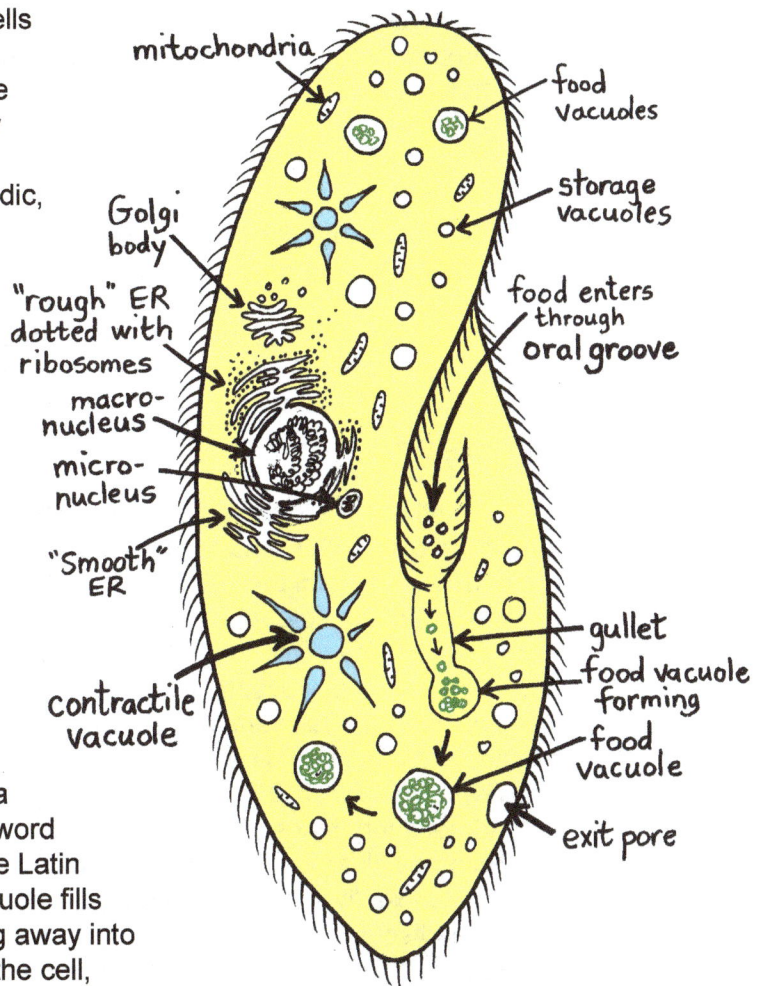

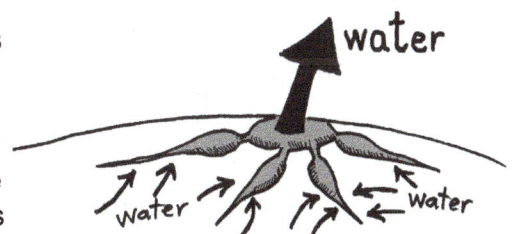

mitochondria

food vacuoles

storage vacuoles

Golgi body

"rough" ER dotted with ribosomes

food enters through oral groove

macro-nucleus

micro-nucleus

"Smooth" ER

contractile vacuole

gullet

food vacuole forming

food vacuole

exit pore

water

water   water

The contractile vaculole contracts (shrinks) suddenly, forcing water out the circular part in the middle.

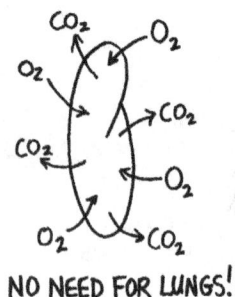
NO NEED FOR LUNGS!

A Paramecium needs oxygen, just like you do. You require lungs to get oxygen from your environment. A Paramecium can simply take oxygen right out of the water around it. It doesn't even need gills like a fish does. Oxygen molecules are so small that they can leak through the membrane and the pellicle. (When tiny particles go across a barrier, this is called *diffusion*.) Oxygen diffuses into the Paramecium and floats around in the cytoplasm. Like all of our cells, the Paramecium makes carbon dioxide as a waste product. We breathe out carbon dioxide when we exhale. In a Paramecium, the carbon dioxide goes right through the outer membrane, out into the surrounding water.

A Paramecium does not need lungs, nor does it need a heart or a system of veins and arteries. We need these things because we are large organisms and have many cells deep inside our bodies, far away from sources of oxygen. Oxygen must be transported to all of our cells. A Paramecium is so small that diffusion is good enough—no need for any transportation!

Four other structures found in all animal, plant and protozoan cells are mitochondria, ribosomes, endoplasmic reticulum (ER) and Golgi bodies. The *mitochondria* are the energy-makers of the cell. They take sugar and oxygen and turn them into energy in the form of ATPs. *Ribosomes* are like little factories. They assemble all the parts and products that the cell needs. (Cell parts are made out of proteins.) Ribosomes are often found near networks of tubes called *endoplasmic reticulum (ER)*. These tubes can both make things and transport things. Some ER looks "rough" because of the ribosomes all around it. If the ER is not dotted with ribosomes, it looks "smooth." The *Golgi bodies* act like little packaging and sorting warehouses, or maybe post offices. They take the proteins made by the ribosomes and make sure they get delivered to places where they are needed.

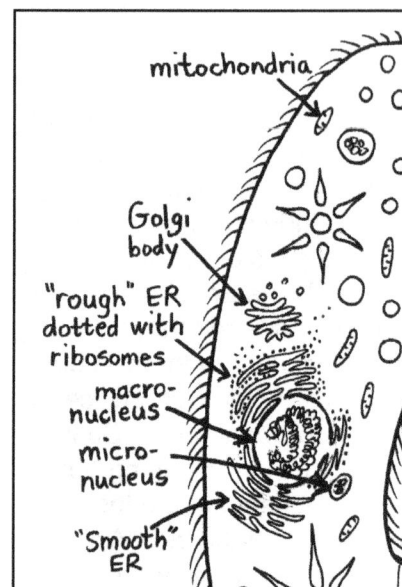

If you want to learn a lot more about these cell parts, try the curriculum called Cells by the author of this book. You can find it on Amazon.com, or by going to www.ellenjmchenry.com. For those of you who have already read Cells, you may be wondering if the Paramecium has lysosomes and a cytoskeleton. Yes, it does, but information about these is extremely difficult to find. You have to read post-graduate-level research papers, and even these are very few and far between. There just isn't a lot of information about the normal organelles of a Paramecium.

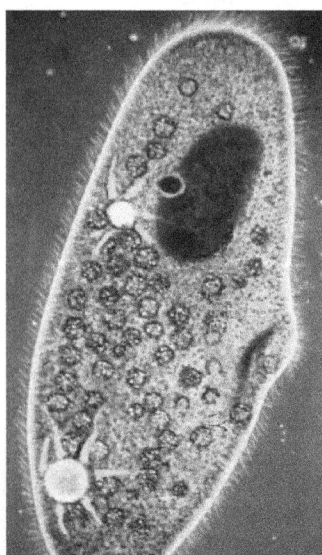

The endoplasmic reticulum is connected to the *nucleus*. All cells have a nucleus. This is where DNA is kept. The DNA is like a library that contains all the information a cell needs. It has instructions for how to make and repair all the cell's parts. Most cells, including yours, have only one large nucleus. A Paramecium has two nuclei: a large *macronucleus* and a small *micronucleus*. You can see them in this photograph as the large and small black ovals. The macronucleus is absolutely stuffed full of DNA. For unknown reasons, it has lots and lots of copies of the same information. It's like having multiple copies of the same book in a library. In some cases a macronucleus can have as many as 800 copies of the same information. Why? There must be a reason, but researchers have not discovered it yet.

The micronucleus has all the same information as the macronucleus, but it does not have lots of extra copies, so it can be much smaller. The micronucleus doesn't seem to do anything while the cell is going about its normal daily life. Its only job seems to be allowing the cell to exchange DNA with another Paramecium before it splits in half.

Paramecia aren't male or female. They have no gender at all. You can't say "he" or "she" for a Paramecium; you have to say "it." They have no reproductive parts, so they don't lay eggs or have babies. The methods they use for reproduction are very basic.

A Paramecium can reproduce in two ways:

1) It can simply split in half (making clones of itself), or

2) it can exchange DNA with another Paramecium before it divides, thus creating new Paramecia that are not identical to the original (a bit more like children instead of clones).

The first method—just splitting in half—is called **binary fission**. (Binary comes from the Latin root word "bini" meaning "by twos." Fission is a Latin word that means "splitting." So binary fission means splitting in two.) The Paramecium makes extras of its inner parts, then splits in half. The two new cells are exact copies of the original one. Then those two new Paramecia can each split in half. Now we have four. Then those four can split in half, making eight. Eight goes to sixteen, sixteen goes to thirty-two, and so on. This process can go on for quite some time, but scientists have discovered that eventually the clones-of-clones-of-clones-of-clones-of-clones-etc. become weaker and weaker until they are unable to continue dividing. The process of binary fission "wears out" after about 200 splits.

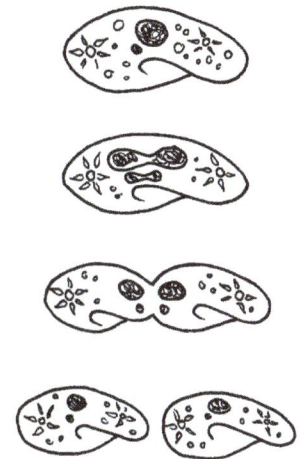

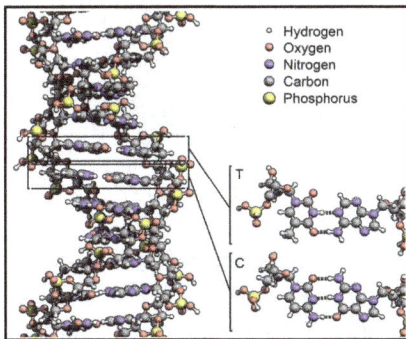

DNA is made of a ladder-shaped string of molecules. Information is stored as a pattern in the rungs of the ladder.

Researchers guess that each time the macronucleus makes a copy of itself, tiny mistakes occur. The mistakes are small, so it does not make a difference at first. But after about 200 duplications of the DNA, so many mistakes have occurred that the information starts to be affected. The Paramecium's instructions for how to make and fix cell parts are now hard to understand or are wrong. The cells begin to be unable to do the processes that they must do to keep themselves alive, so they die. (The picture to the left shows the molecular structure of DNA. If those "beads" get mixed up, the information does, too.)

This is where the micronucleus comes in. The DNA must be fixed. The fixing process involves trading DNA with another Paramecium. The Paramecium somehow knows how to find another Paramecium who also needs to trade DNA. They line up next to each other, side by side, and at the place where they touch, their cell membranes and pellicles dissolve so that cytoplasm can stream back and forth. (The technical term for this process is **conjugation**.) Then the micronucleus of each Paramecium divides several times (using a process called meiosis which we are not going to explain right at this moment because it is a bit complicated).

Diagram of what happens

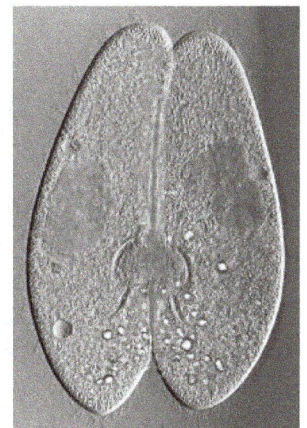

Actual photo

These divisions create little "half nuclei." Each Paramecium gives a "half nucleus" to the other. Then the two "half nuclei" join together to form a new whole. Now each Paramecium has a new micronucleus, half of which came from the other Paramecium.

Right after this happens, the Paramecium's old macronucleus dissolves and disappears forever. The new micronucleus divides and makes a new macronucleus. Thus, the old, defective DNA is gone and the new combination takes over. The new, revitalized Paramecium can then start using binary fission again until it wears out after several hundred divisions.

Paramecia can't bite, but they <u>do</u> have a secret weapon. Attached to the inside of the pellicle are tiny structures called **trichocysts**. ("Tricho" is Greek for "hair" and "cyst" comes from the Greek word "kystis" meaning "bag or pouch." So it's a hair in a bag.) A trichocyst is a bit like a harpoon tied to a string, all curled up inside a hidden pouch, waiting to spring out and pierce whatever happens to be in its line of fire. Scientists are still not totally sure they have figured out all the functions of the trichocysts, but the Paramecium does seem to use them as weapons. They might also be used to anchor the Paramecium to something while it feeds. But using them as weapons is a lot more interesting, so let's investigate that use.

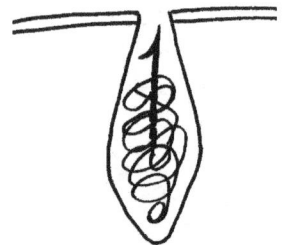

If the Paramecium gets into a situation where it is threatened by a predator, a chemical signal (using calcium) can be released thoughout the cell, causing a very rapid chemical change inside the trichocysts, resulting in the harpoons being launched out at a very high speed. How effective are these little weapons? Well, to be honest, if the predator is at least as large as the Paramecium, the harpoons are far from deadly. The predator will likely be back soon for a second try. In the meantime, though, the Paramecium will have turned and started swimming as fast as it can in the opposite direction. Retreat is one of the Paramecium's primary survial strategies.

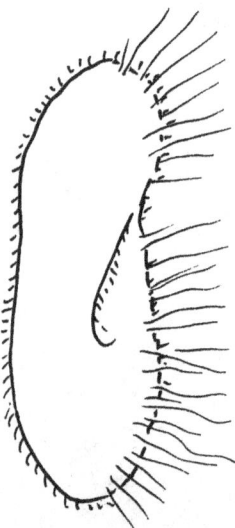

Unlike sailors on old-fashioned whaling boats, the Paramecium can't reel in its harpoons and reuse them. Once they are fired, that's it. It would take muscles to pull the harpoons back, and the Paramecium does not have msucles. Instead, the cell machinery (the ribosomes, the endoplasmic reticulum, the Golgi bodies, and other parts) gets busy making a new batch of trichocysts. The old ones are dissolved and the new ones take their place. This process must go on continually for the Paramecium to stayed armed.

Trichocysts are very small. A Paramecium has hundreds, or perhaps thousands, of them, hiding among the cilia. Because they are so small, and so numerous, they were not shown on the big diagram on page 9.

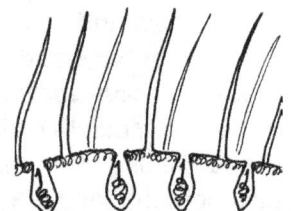

Trichocysts can fire on just one side.

ACTIVITY 1.1   Watch some videos of real Paramecia

Go to the special channel that was set up for this curriculum at: YouTube.com/TheBasementWorkshop. Click on "Playlists," then on "Protozoa." The videos that go with this chapter will be labeled as Chapter 1. You'll see Paramecia swimming, eating, and even firing off their trichocysts.

ACTIVITY 1.2   Comparative anatomy

Can you match the Paramecium part with its equivalent human part? Draw a line between the matches.

Paramecium parts:
- pellicle
- cilia
- oral groove
- gullet
- anterior
- contractile vacuole
- food vacuole
- trichocysts
- no equivalent

Human parts:
- mouth
- head
- throat
- kidneys
- stomach
- lungs
- no equivalent
- arms and legs
- skin

ACTIVITY 1.3   Can you answer these questions?

1) This word means "made of more than one cell." _____

2) The Greek word "paramekes" means _____.

3) For two hundred years, Paramecia were usually called little _____.

4) A Paramecium's thick-yet-flexible outer layer is called the _____.

5) On top of the layer in the previous question, there is a very thin layer called the _____ _____.

6) Where are your eyes and nose located—on your anterior or your posterior? _____

7) A Paramecium does not need lungs because oxygen can simply _____ through its outer layers.

8) Which body part does the Paramecium use to swim? _____

9) Which body part does the Paramecium use to pump out excess water? _____ _____

10) Which body part only becomes active during conjugation? _____

11) Parameciua use this method of reproduction most of the time: _____ _____

12) Approximately how many times can a Paramecium split in half before it must go through conjugation? ___

13) What does a Paramecium eat? _____

14) The Latin word root "or-" means _____.

15) The Greek word "tricho" means _____.

BONUS QUESTION: Which body part is responsible for generating energy? _____
SECOND BONUS QUESTION: Which body part sorts things like a post office does? _____

ACTIVITY 1.4   ZEUS versus JUPITER, round 1

The Titans have fought over many things, but... vocabulary words? Doubtful. Well, then, this will be a first for them! The god with the most words wins. (Actually, Zeus and Jupiter were the same deity because the Romans borrowed Zeus from the Greeks and changed his name to Jupiter, so it's a "win-win" situation. The king of the pantheon wins, either way!)

We have listed Greek and Latin word roots found in words in this chapter that are printed in bold italic type (the words that are darker than the others). The Greek word roots are under Zeus, the king of the Greek gods, and the Latin word roots are under Jupiter, the king of the Roman gods.

In this first chapter, it looks like Jupiter wins by three words.

Your job in this activity is to figure out which vocabulary words contain these word roots. Look back through the chapter and find those words in bold type. Write the appropriate vocabulary word (or words) on each line. (For example, the word "microscope" contains two Greek words: "mikros" and "skopos." So write the word "microscope" on the line after "mikros," and also on the line after "skopos.")

GREEK

mikros (small) _____

makros (large) _____

metron (measure) _____

protos (first) _____

zoion (animal) _____

skopos (to watch) _____

paramekes (oblong) _____

kystis (bag) _____

kytos (container) _____

tricho (hair) _____

plasma (plastic) _____

soma (body) _____

poros (passageway) _____

endon (inside) _____

mitos (thread) _____

chondrion (small grain) _____

LATIN

multus (many) _____

uni (one) _____

bi (two) _____

ante (before) _____

post (after) _____

pellicula (husk) _____

cilia (hair) _____

or (mouth) _____

vacuus (empty) _____

ex (out) _____

porus (passageway) _____

fissus (split) _____

con (with) _____

jugum (yoke) _____

mille (1,000) _____

nucula (little nut) _____

diffusio (to spread out) _____

gula (throat) _____

membrana (thin parchment) _____

14

CHAPTER TWO: MORE CILIATES

Meet **Didinium**, the tiger of the protozoan world. It's a **carnivore**. ("Carn" comes from Latin and means "flesh or meat." "Vore" is from the Latin word "vorare," meaning " to eat.") Didinium kills and eats other ciliates, even some twice its size, and its favorite meal is a Paramecium. Why it prefers Paramecium to other ciliates is unknown. It can be forced to eat other things, but if Paramecia are around, it will go after them first.

Didinia (the plural of Didinium) might look very different from Paramecia, but they actually are quite similar. Almost everything we learned about Paramecia is true for Didinia. They have the same cells parts, they swim through the water using cilia, they reproduce using binary fission and conjugation, they have food vacuoles and exit pores, and they can fire little harpoon darts.

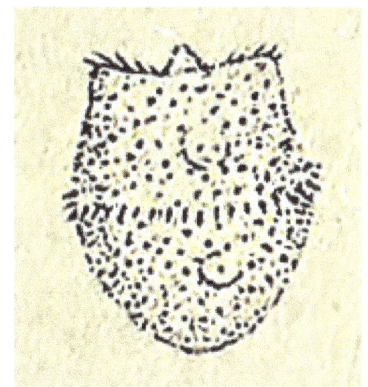

Drawing by Otto Müller, 1786

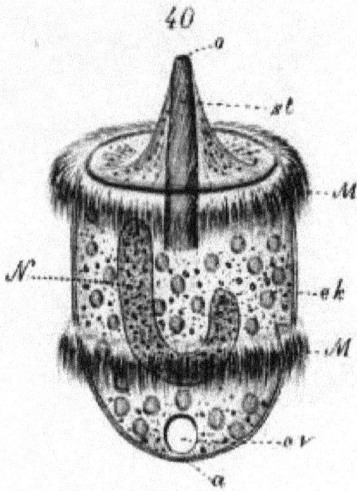

A Didinium drawing from 1896

The two biggest differences between Paramecia and Didinia are the arrangement of the cilia and the shape of the "mouth." The Didinium's cilia are arranged in two neat rows, one along its "belt line" and one around the top rim. When it beats these cilia, it spins through the water. The thing sticking out of the top of Didinium is called the **cytostome** and is equivalent to the Paramecium's oral groove. It is the opening where the food goes in. (Cytostome is Greek for "cell mouth.") Yes, that cytostome really stretches! Can you imagine it taking in an animal as large as the Didinium itself?

That dark "J" in the middle is the macronucleus. All those other little dots are probably food vacuoles. You can see the exit pore at the bottom. When this picture was drawn, over 100 years ago, scientists had no idea what the organelles did. They did not yet know about DNA, so they really had no clue what the role of the nucleus was.

The Didinium makes specialized trichocysts called **toxicysts**, which contain a toxic (poisonous) substance. The toxic chemicals are capable of paralyzing the Didinium's prey. This is part of the reason Didinium is able to capture prey that is as large, or larger, than itself.

These drawings are from 1909.

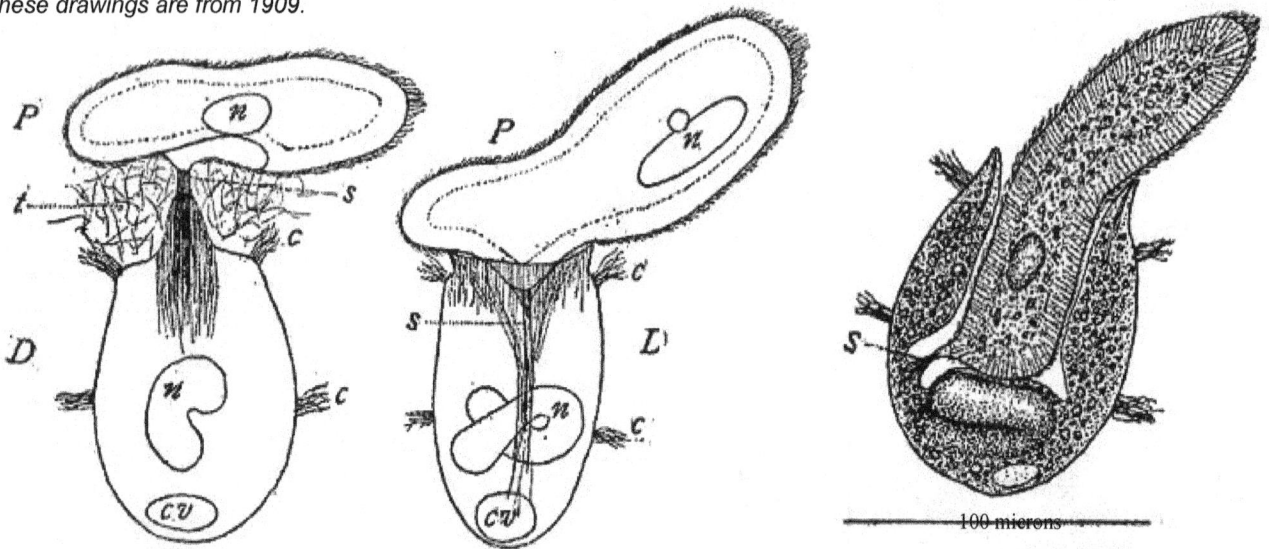

Didinium "stings" the Paramecium with its toxicyst darts.

After the Paramecium is still, Didinium is able to start ingesting it.

100 microns

A cross section of a Didinium with a half-eaten Paramecium. (Notice the line that is 100 microns long.)

Aah!! It's a little sea monster!

Oh, no! I wonder if it's a carnivore, too?!

Look! It just clobbered the didinium!

And zapped it, too. Poor didinium!

I thought you were mad at didinium for trying to eat the paramecium.

I always cheer for the "under dog."

Ooo-yuck! Didinium's guts are spilling out everywhere! It exploded!

Meet **Dileptus**, the elephant of the protozoan world. (Killer elephant, that is.) Dileptus reminds many people of an elephant because it is large and has a long **proboscis** at its anterior end. (The word proboscis comes from the Greek words "pro" meaning "for," and "boskein" meaning "to feed." The Greeks called an elephant's trunk a "pro-boskis" meaning "for feeding.") But you may think Dileptus looks like something else — a plesiosaurus with no fins, a leg-less swan, a snake that ate a watermelon?

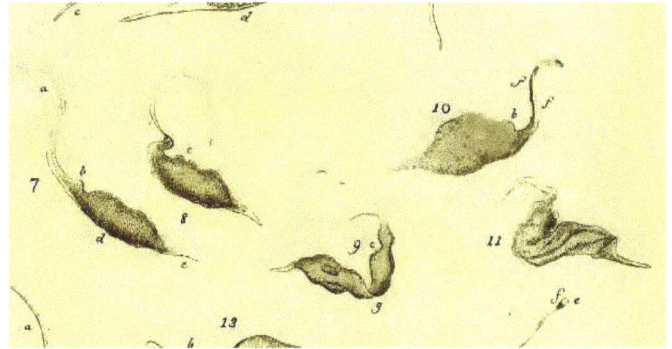

Drawing by Otto Muller, 1786. He called them "vibrios."

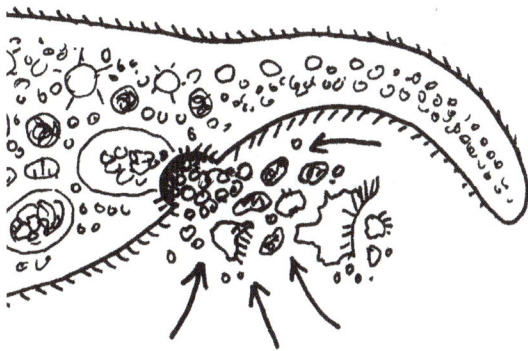

Dileptus can't use its proboscis for feeding in the same way an elephant does. An elephant uses its trunk almost like we use our arms, hands and fingers. Dileptus is not nearly as smart as an elephant, but it does pretty well for a single cell. It uses its proboscis to injure its prey. Just like Didinium, Dileptus has toxicysts (tiny harpoons with poison tips). These toxicysts are located primarily at the base of the proboscis. A toxic smack from a Dileptus proboscis can completely rupture the pellicle of most ciliates. As the prey's cytoplasm and organelles ooze out into the water, Dileptus starts sucking them up with its "mouth" (cytostome) which is located at the base of the proboscis. Cilia help to sweep the particles toward the cytosome.

The Dileptus forms food vacuoles, just like Paramecia do. The food vacuoles float around the cell as digestive chemicals go to work breaking down the food particles into smaller and smaller pieces. Eventually, the protein, fat, and carbohydrate molecules inside the food vacuoles are released into the cytoplasm where they can be used as raw ingredients for manufacturing new cell parts for the Dileptus. (If the Dileptus then gets eaten by a larger predator, the Dileptus' cell parts will be recycled and used by the animal that ate it. And if that animal then gets eaten, its parts are recycled. And so it goes, up the food chain, until it reaches the very largest predator who does not get eaten. But that predator dies eventually, and then bacteria do the recycling.)

There are about a dozen different types of Dilepti (one Dileptus, two Dilepti). They are all similar in shape, but not identical. This picture shows shapes from actual photographs. What animals to they remind you of? We think the first one looks like a seahorse. The second one (on top) looks like a seal, or maybe a baby bird with a long tail. Seriously, though, can you tell which of these Dilepti just ate a big meal?

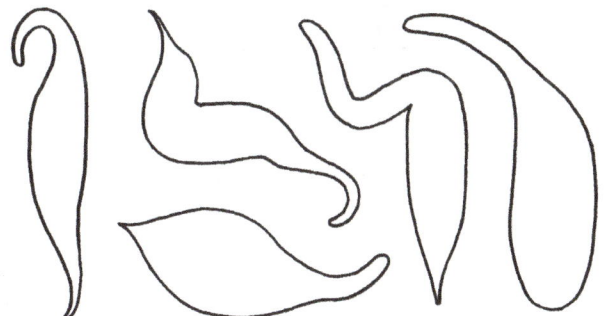

17

Dileptus looks as if it is filled with tiny circles, like marbles of various sizes. Many of these circles are vacuoles of various types: food vacuoles, storage vacuoles, and contractile vacuoles. Paramecium and Didinium had just a few contractile vacuoles. Dileptus has many more. Most of the contractile vacuoles are lined up in row, along what would be its back if it were a larger animal. Since Dileptus is only a single cell and doesn't really have a proper back, that side is simply called the **dorsal** side, meaning the side opposite the mouth opening. (The word "dorsal" shows up a lot when you study animals and it always means something having to do with the back. Dorsal comes from the Latin word "dorsum" meaning "back.")

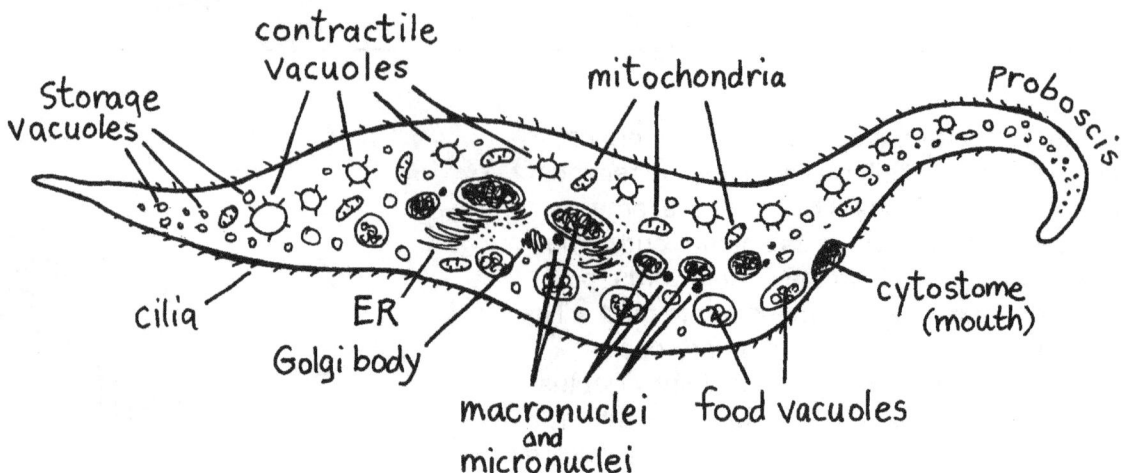

Another notable difference between a Dileptus and a Paramecium is the number of nuclei. A Paramecium has just one big macronucleus and one small micronucleus. A Dileptus can have up to 60 or 70 macronuclei and hundreds of micronuclei. Why it has so many is still a mystery. Do all of the micronuclei play a role in conjugation (trading DNA), as we saw in the Paramecium? We know that in some ciliates with multiple nuclei, more than one micronucleus is traded back and forth during conjugation. But no one has ever been able to observe Dilepti for long enough to determine what happens to all of their micronuclei. (It's important to remember that scientists don't know everything!)

The Dileptus is not the largest ciliate in the pond. There are several well-known ciliates that can grow to be much larger than a Dileptus. In fact, they are so large that you can see them without a microscope if you have really good vision. You won't see any detail, of course. They'll just look like a little dot smaller than the head of a pin.

Meet **Spirostomum**, the mammoth water snake of the protozoan world. (Its name comes from the Greek words "spiros" meaning "spiral," and "stomum" meaning "mouth.") Actually, the Spirostomum doesn't have much in common with snakes of any kind—it doesn't bite, poison, or squeeze its prey. It's much less aggressive than Didinium or Dileptus and does not have trichocysts. It's more like a hungry swimming tree trunk. Its lifestyle is fairly similar to that of the Paramecium; it wanders about taking in whatever bits of food it can find: bacteria, algae, tiny pieces of rotting plants, and small protozoa.

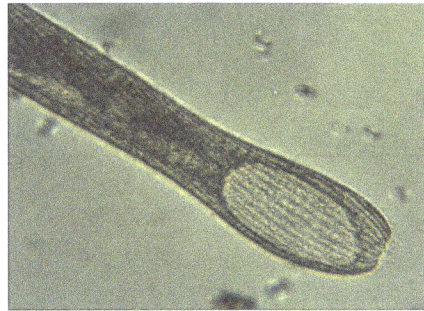

The most notable feature of the spirostomum is the stripy pattern on one end. This is the posterior end (the "tail"). The stripes are sort of like muscle fibers and they run the whole length of the animal even though you can only see them in this posterior region. (They are called *myonemes*, for those of you who like to know the proper names for things.) The reason you can see them here is that at the posterior end is a giant vacuole (a contractile vacuole that expels water). This vacuole is fairly transparent ("see-through") so you can see the myonemes.

The cytostome ("mouth") is very hard to see because it is simply a small slit in the anterior end. It is similar to the oral groove and gullet of the Paramecium. The cytostome does not stretch like the Didinium's, so the Spirostomum can't possibly eat anything super large.

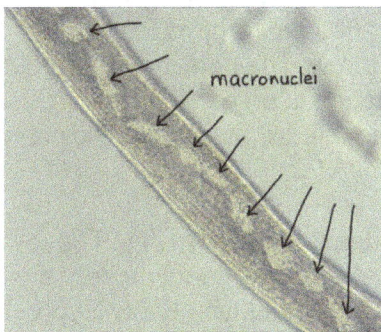

A very noticeable feature inside the Spirostomum is the string of macronuclei, all joined together like a string of beads. Some scientists say it is just one really long nucleus pinched off into smaller pieces. It makes sense for the Spirostomum to have a very long nucleus because, as you will remember, it is the DNA in the nucleus that provides information to the cell's organelles. If the nucleus was only at one end, the poor organelles at the other end would be far away from the source of information, making it very hard for them to do their jobs.

The Spirostomum uses binary fission as its primary method of reproduction. Like all ciliates, it also uses conjugation once in a while, whenever it feels its DNA is getting "worn out" and needs to be traded. Trading DNA revitalizes and renews them.

The Spirostomum does not usually wriggle back and forth like a snake. It glides smoothly through the water using rows of cilia so small that you can't see them without very high magnification. It can change direction easily and quickly without turning around. Spirostomum starts its cilia going in the opposite direction and it moves backwards. Even though it usually stays straight, it can bend if it needs to and is capable of making a U shape.

Spirostomum's claim to fame is that it can contract its body down to about one fourth of its original size in a fraction of a second (1/200 of a second). No other living cell can contract faster than a Spirostomum. The contraction happens whenever the Spirostomum is startled. In the lab, you can make them contract by gently tapping the microscope slide.

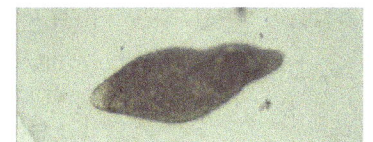

Is the Spirostomum the biggest protozoa in the pond? Turn the page and find out.

Meet **Loxodes** (*locks-OH-dees*), the whale of the protozoan world. It's most similar to a baleen whale because although it is very large it eats small things. Baleen whales eat a diet of microscopic plankton. Loxodes eats bacteria, algae, and tiny protozoa. Those large circles you see in the drawing of Loxodes are food vacuoles filled with tiny food particles. Those guys in the boat aren't in much danger of being eaten. They're probably too large.

Loxodes is a ciliate, just like the others we've met, so you already know all the basics about it — how it swims (cilia), how it digests (food vacuoles), and how it reproduces (binary fission and conjugation). Loxodes is different from other ciliates in two respects, however. The first is that it does not have any contractile vacuoles. The second is that it has a special type of organelle that helps it know which way is "up." You can see a string of these organelles in the picture above. They look like perfectly round fried eggs lined up in a row. They are called **Müller bodies**, named after Otto Müller, the guy from the 1800s that we met back on page 6 when he wanted to spell Paramecium with an "o" (paramoecium). He also did the drawing of the Didinium on page 15 and the drawing of the Dilepti on page 17.

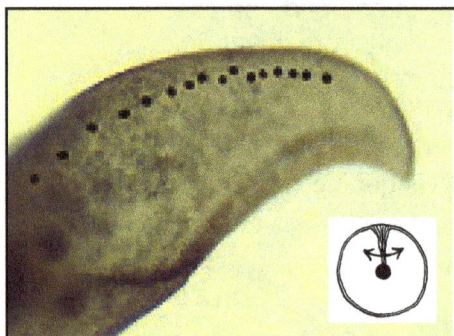

The Müller bodies allow the Loxodes to sense gravity. They work a little bit like the cells in your inner ear, which let you sense the tilt of your head. Even with your eyes closed, you know if your head is level or is tipped forward or backward. The little inset picture shows how the dark blob in the middle of the organelle is suspended by a thin protein thread. It's like a ball on a string. Imagine rotating that circle clockwise. What would happen to the position of the ball? It would go toward one side, wouldn't it? When the inner part of the Müller body migrates toward one side of the circle, chemical and electrical changes take place around it.

Why would Loxodes need to sense gravity? The best guess so far is that it needs to migrate to areas of less oxygen. Loxodes doesn't like too much oxygen. When there is too much oxygen near the surface, Loxodes takes a dive.

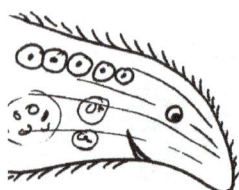

It's easy to tell which end of Loxodes is the anterior end, since it is pointy-looking, sort of like a beak. It is called the **rostrum**, which is Latin for "beak." It's not really a beak, of course. It just reminds us of one because of its general shape. The mouth opening is very small and is located right under the rostrum, just where you would put the mouth if you were making a cartoon Loxodes. (We added a cartoon eye, too!)

So... is Loxodes the largest ciliate? (Okay, okay, you've already looked at the next page.)

Panel 1: I'm kind of afraid to ask this, but how much bigger can these critters get?

Panel 2: Hmm... It looks like there is one that can grow to be 2-3 mm. And how big is our boat?

Panel 3: I think we figured out that it's 1000μm which is 1 mm. So it's 2 to 3 times the length of the boat.

Panel 4: And its cilia can create a whirlpool to draw in prey. I think my end of the boat is sinking...

Panel 5: A whirlpool ?!

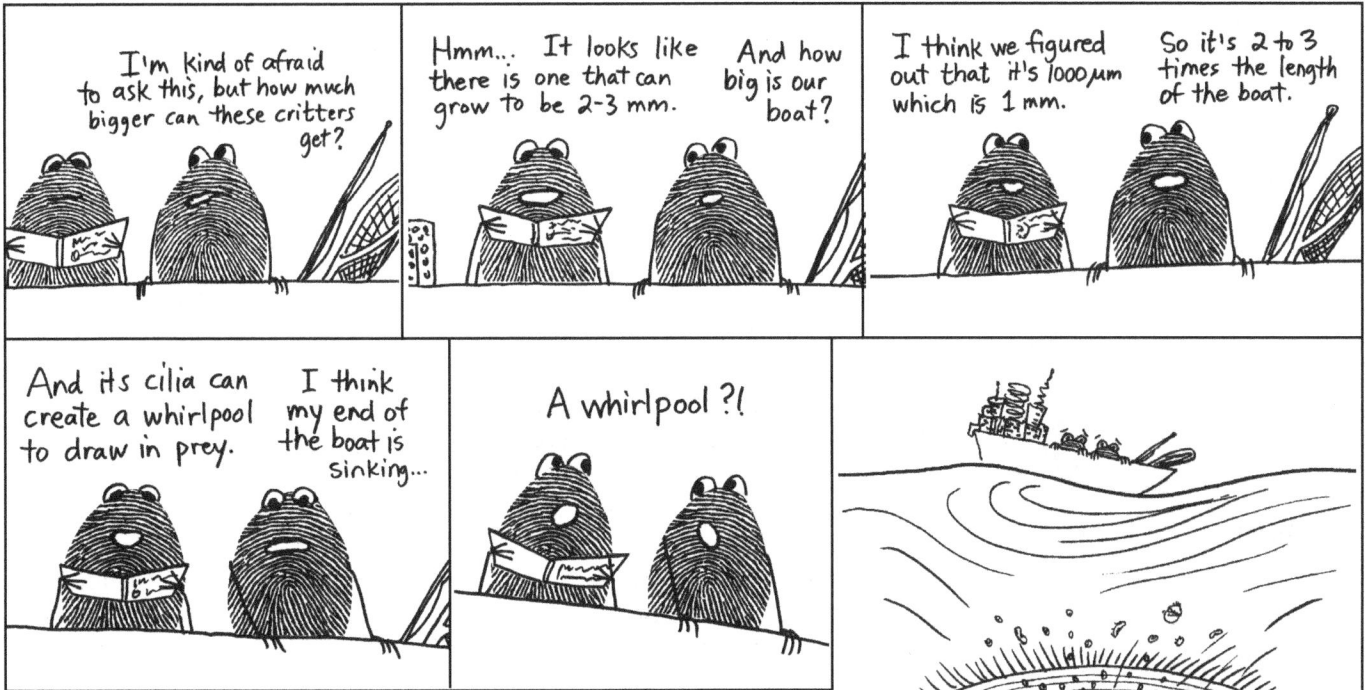

Meet **Stentor,** the living trumpet. It is named after an ancient Greek who was known for his loud voice. He was a herald (announcer) in the Greek army. Supposedly, the voice of Stentor was louder than the combined voices of 50 men. Things were fine as long as he remained humble about his ability. However, as so often happened in these legends, the mortal got overly confident about his abilities and challenged one of the Olympian gods. Stentor foolishly agreed to a shouting contest with Hermes, the herald of the gods. Stentor lost not only his voice, but his life.

A simple device that can be used to magnify the sound of your voice (to make you as loud as Stentor was) is a cone-shaped device called a megaphone. The protozoan Stentor is the same shape as a megaphone.

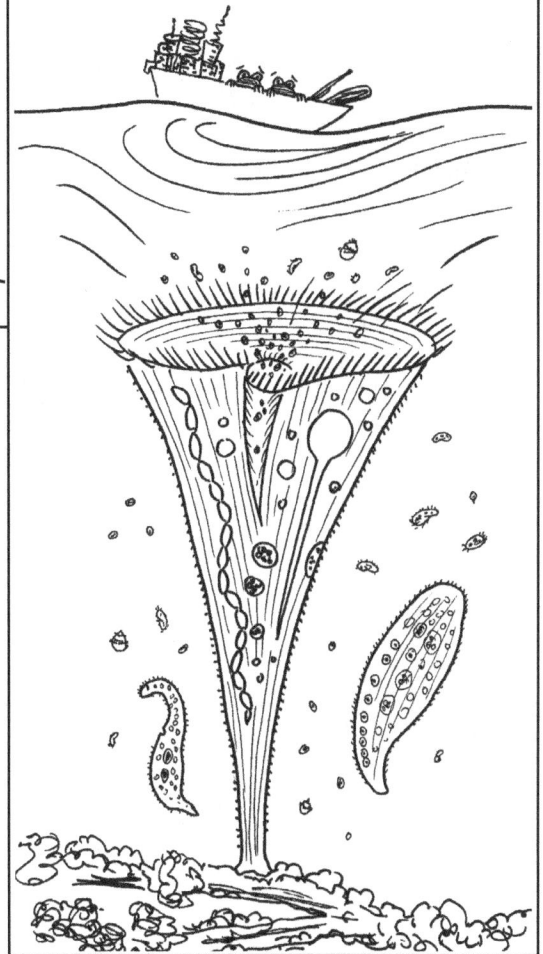

The Stentor is a ciliate, just like all the protozoa we've met so far. In the next chapter, we will meet some protozoa who are not ciliates, but so far everyone we've met has cilia. The Stentor has cilia on its body and also along the top rim of its "trumpet." The cilia along its body are shorter and are used for swimming. The cilia on the rim are longer and are used for creating a vortex (whirlpool) in the water, which draws food down. Most often you will find Stentor standing still. (The fancy word for this is **sessile** (seh-sill), which means "attached to something and not swimming around"). The Stentor can attach to any large piece of floating debris, like a twig or leaf, or a floating clump of algae. It stays there while the supply of food lasts. When it senses that the food supply is dwindling, it detaches itself and goes off to find a new place to attach.

You can see that Stentor has the same kind of nucleus as Spirostomum. Its looks like a long string of oval beads. Long protozoa need long nuclei! The blank circle (with a long, thin tube running down from it) is a contractile vacuole. The Stentor is continually taking in water and needs a way to pump it back out again. Every few minutes the contractile vacuole reaches capacity and gives a sudden and strong "squeeze" to expel the water. (The long tube part helps to collect water from the lower part of the cell.) You can also see lots of food vacuoles.

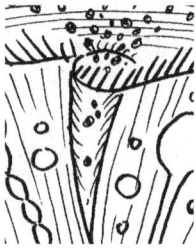

The V-shaped thing in the middle of the top is the "mouth." The proper name for it is the **buccal cavity.** (You say the word "buccal" exactly like the word "buckle." So next time you are required to think of homonyms for an English assignment, you can use this one and amaze everyone with your knowledge of biology!) The word "buccal" comes from the Latin word "bucca" meaning "mouth" or "cheek." The muscle in your cheek (which lets you draw in your cheeks to make the silly "kissing fish" face) is called the buccinator muscle (*buck-sin-a-tor*).

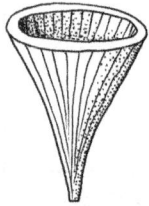 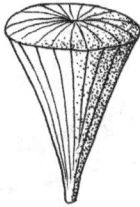

WRONG   RIGHT

As we look at the buccal cavity on the top of Stentor, we need to understand something very important about the overall shape of the Stentor. It is very easy to misinterpret drawings and photographs. Most students get the impression that Stentor is shaped like a hollow cone, like the drawing on the left. We must assume that the organelles are somehow embedded in the walls. In reality, the Stentor is not hollow, but has a top. It is like a full ice cream cone, not an empty one.

This antique drawing (from over 100 years ago) is perhaps the best drawing you can find on the Internet. It is the drawing that most clearly shows that the top is solid, not open. Notice the buccal cavity going into the cell like a funnel. The contractile vacuole and the macronucleus look very much like the ones in our cartoon picture on the previous page.

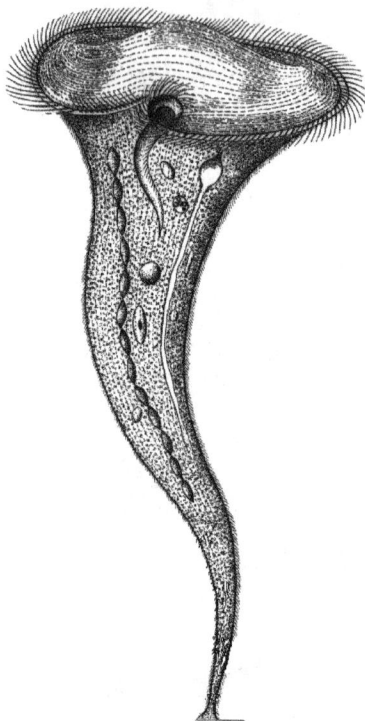

Again, don't forget that protozoa are transparent (see-through). The nucleus (even in this very good picture) looks like it is pasted on the side. It is actually *inside* the cell, not on the surface.

The skinny part at the bottom is called the **peduncle** (*PED-uncle*, or *peh-DUNK-el*, take your pick). This word comes from the Latin root "ped" meaning "foot," and the Latin ending "unculus" meaning "little." (As a brief side note for those of you interested in Latin, this word ending doesn't mean "small" in the same way that "brevis" and "parvus" do. Rather, "unculus" is similar to what we do with "y" in English. A grown man is "Bob," and his little boy is "Bobby.")

A Stentor attaches to things using the peduncle. It can secrete a sticky substance that will hold it in place.

Besides being the largest of the ciliates, the Stentor has two other characteristics that set it apart from most other protozoa. First, it can regenerate if it is chopped into pieces. Each piece will grow into a new Stentor. Some researches claim they have grown Stentors from pieces as small as 1/100 of an original cell. This is not only amazing, but technically impossible. Why? Remember that the cell only "knows" what to do because of the information stored in its DNA in the nucleus. If it needs new cell parts, the information for how to make them comes from the DNA. A piece of Stentor that had a piece of nucleus would have the necessary information to be able to grow new organelles. But even if you separate that string of macronuclei into individual beads and give one to each piece, you still would not have enough for 100 pieces. How do the pieces without any DNA know what to do? Are there bits and pieces of DNA floating around in the cell? (If you become a cell researcher someday, here is a puzzle for you to solve!)

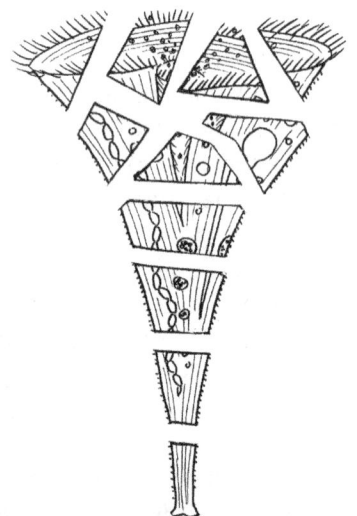

The second somewhat unusual thing Stentor can do is to allow algae to live and grow inside its body without digesting them. This is what gives green Stentors their color. It's the algae that are green, not the Stentor.

The Stentor takes in algae through the buccal cavity and puts them into food vacuoles. But somehow or other the algae resist digestion and eventually escape from the vacuoles. They float around in the Stentor's cytoplasm, like fish in an aquarium. The Stentor doesn't mind a bit; they don't cause any harm. In fact, they are somewhat helpful. The algae can take carbon dioxide, one of the waste products that all living cells make, and use it for photosynthesis, making sugars that both the algae and the Stentor can eat. So the algae sort of function as both housekeepers and cooks!

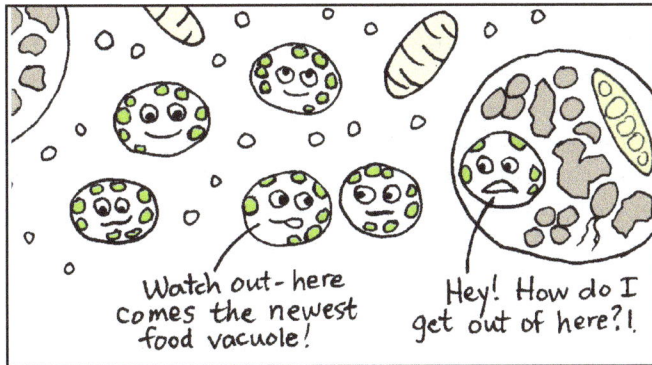

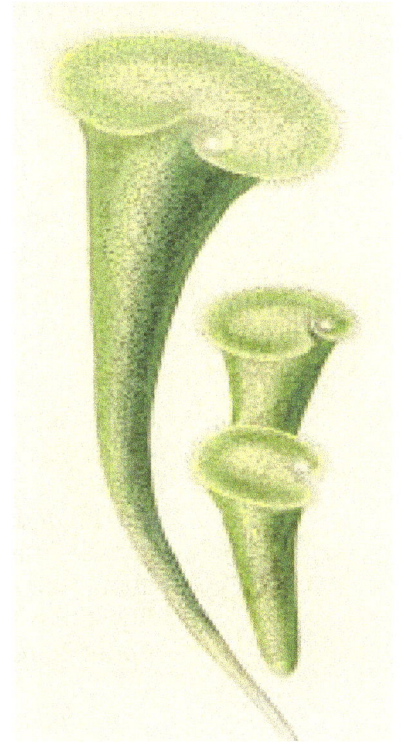

Watch out- here comes the newest food vacuole!

Hey! How do I get out of here?!

A protozoan that is very similar to Stentor is Vorticella *(vort-i-SELL-ah)*. If you stretched a Stentor's peduncle so that it got very skinny, and you rounded the trumpet into more of a bell shape, you'd have a Vorticella. They are considered to be closely related.

As with Stentor, when you look at drawings of Vorticella you often get the wrong impression of its shape. Though it is shaped like an upside down bell, it is not hollow. The SEM photograph on the right shows the true shape of the top of the bell. You can't see this when you look at pictures taken with a regular light microscope

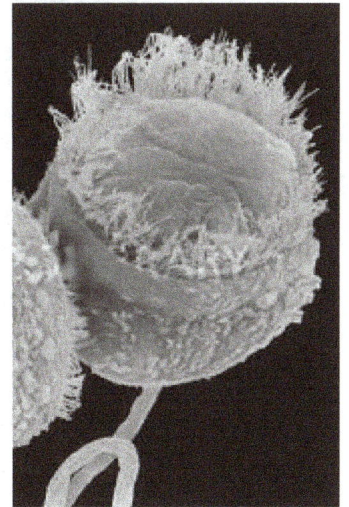

The top of Vorticella's bell is covered; the bell is not hollow.

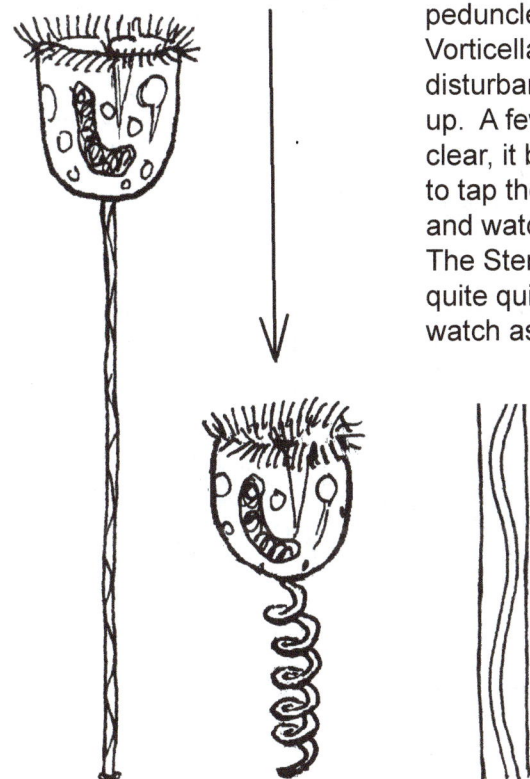

Vorticella's claim to fame is its springy peduncle. It is quite entertaining to watch a Vorticella under a microscope. The slightest disturbance startles it, and it suddenly coils up. A few seconds later, when the coast is clear, it begins to stretch out again. It's fun to tap the microscope slide again and again and watch Vorticella bounce up and down! The Stentor, also, can contract its peduncle quite quickly, but it is not as entertaining to watch as the Vorticella's "spring."

The mechanism that contracts both Stentor and Vorticella is very similar to the one found in Spirostomum. It is called a **spasmoneme** because it causes a very fast contraction, or "spasm." The Vorticella's spasmoneme is visible inside the stalk of the peduncle.

The Vorticella's peduncle is a bit fragile; it doesn't take much to rip it off. The Vorticella can function well enough without it, though. It can start swimming around, like Stentor can. Then it can grow a new peduncle in less than an hour.

Several times we've mentioned "smaller protozoa" that get eaten (along with bacteria and algae) by the larger ciliates such as Spirostomum, Loxodes, and Stentor. Let's wrap up this chapter with a look at the smallest members of the ciliate family.

In descending size, our "Top Four" list is:

Euplotes (yu-PLOH-tees): A unique ciliate that not only swims, but "walks." ("Eu" is Greek for "good" and "plos" is Greek for "swimmer.") It is also a picky eater and spends quite a bit of time sorting out the particles that come into its buccal cavity. It spits out things it does not like. *About 150 µm*

Colpidium (cole-PID-ee-um): A very common ciliate found in most ponds. It looks like a junior version of Paramecium. The name comes from the Greek "kolpos" meaning "gulf." Perhaps Copidium's oral groove area reminded early scientists of a gulf or bay. *About 60 µm*

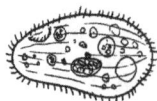

Tetrahymena (tet-rah-HIE-men-ah): Another extremely common ciliate found in many ponds. This ciliate is a favorite of professional cell researchers. Some major discoveries about how cells work were made by studying this ciliate. ("Tetra" is Greek for "four," and "hymen" is Greek for "membrane.") *About 50 µm*

Halteria (hal-TARE-ee-ah): Perhaps the smallest ciliate of all. It is known for its swimming speed and its jumping motion. It was named after the weights used by the Greeks in their version of the long jump event. The weights were called "halteres" and were used to counterbalance body weight during the jump. *About 30 µm*

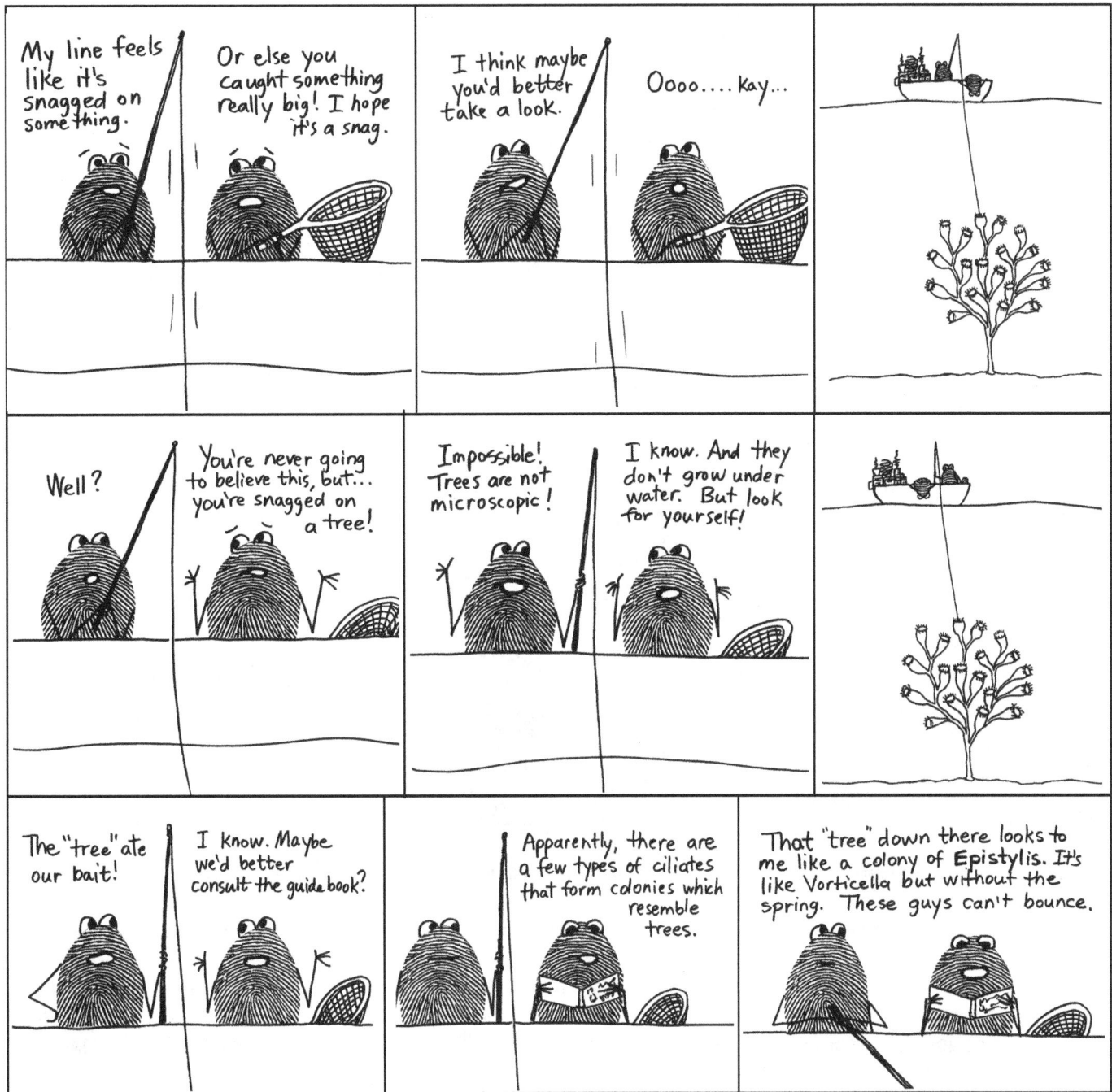

There's one last type of ciliate we really must mention because they are so different from the others. These ciliates form a group called the "Sessilida" because they are **sessile**. (Remember meeting this word back on the page with the shouting Stentor? It is pronounced *(seh-sill)* and it means "can't move around.")

The particular member of this group that our heroes have discovered here is called **Epistylis**. You can pronounce it *(ee-PIST-ill-is)* or *(EP-ee-STY-liss)*. It is almost impossible to find a book or website that tells you how to pronounce this word, so choose one of these and go with it. Its name comes from two Greek words: "epi" meaning "upon," and "stulos" meaning "pillar." The cups are attached to "pillars" and can't move.

These ciliates are very much like Vorticella, except that they do not have myoneme fibers. The myonemes are the muscle-like fibers that allow Vorticella, Stentor and Spirostomum to contract quickly. The Epistylis lacks these fibers, so it has to find another reason to be special.

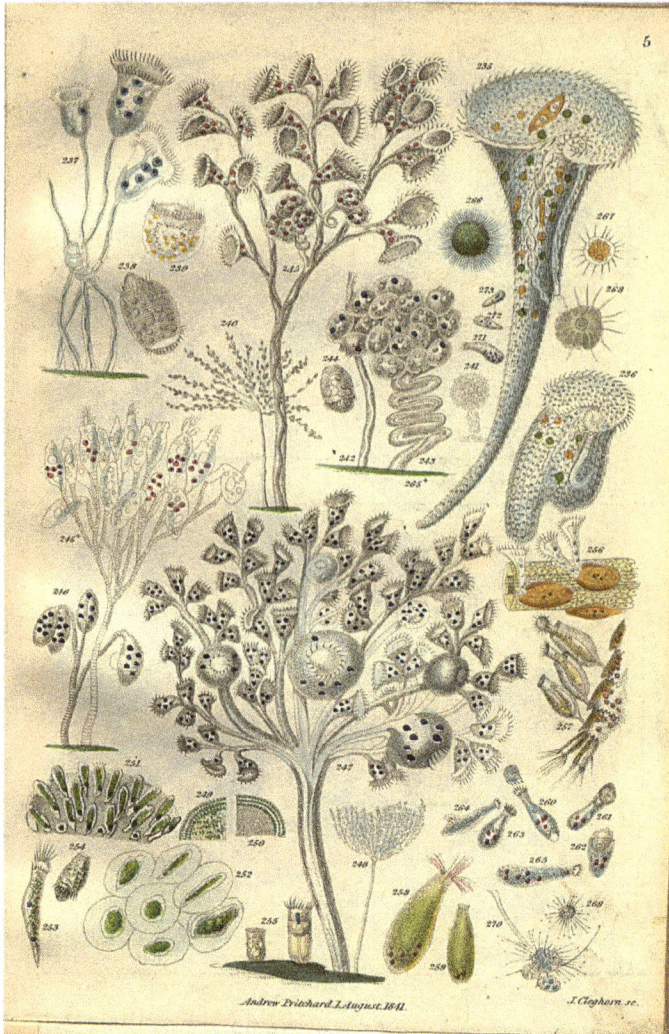

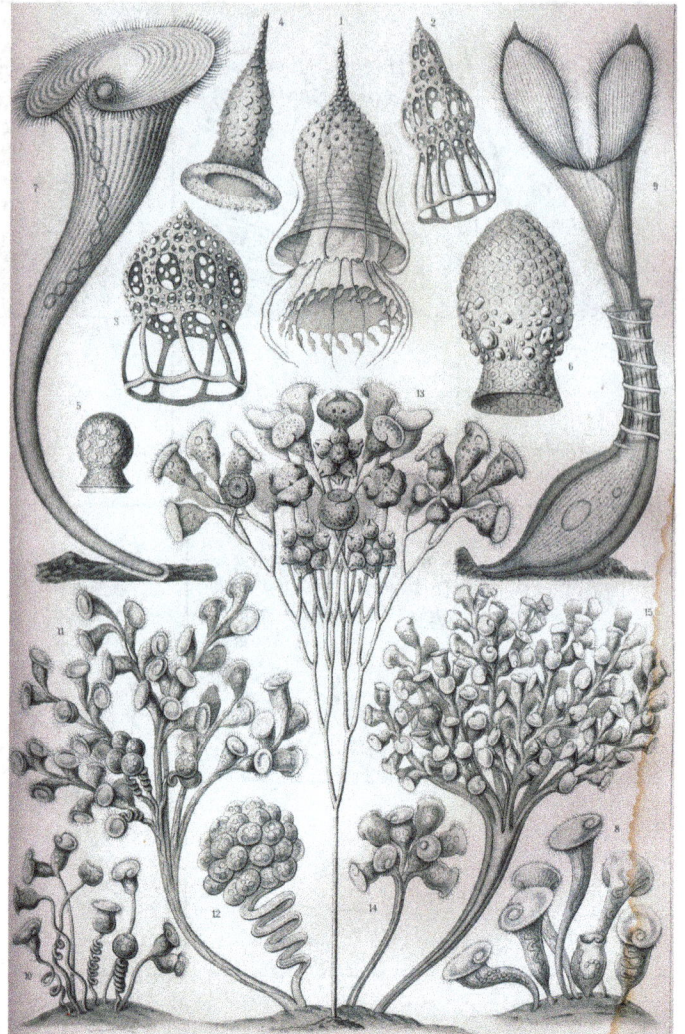

These illustrations were drawn in the mid-1800s. They show ciliates in the group to which Stentor, Vorticella and Epistylis belong.

Epistylis is a single-celled organism so it is capable of living on its own. Each cell is a complete unit, with its own gullet, nucleus (or nuclei), contractile vacuoles, food vacuoles, and other organelles. Technically, it does not NEED other cells. However, it displays *colonial* behavior. This means forming protozoan "neighborhoods." The Epistylis cells permanently bond to each other, forming an overall tree-like shape. Why they do this is unknown. There must be an advantage to living this way or they would not do it.

These beautiful illustrations were drawn in the mid-1800s. Scientists had named and classified thousands of protozoans before the year 1900. Regular "light" microscopes were pretty good back in the 1800s! At 100x or 400x, they were just as good as the microscopes we use today. The scientists of the 1800s could see all the organelles and observe the contractile vacuoles squeezing and refilling. They saw the cilia and the trichocysts. They watched the protozoa go through binary fission and conjugation. They knew what each species ate, and what ate them. What they could not possibly know was the structure and function of DNA, and the molecular processes that go on inside the organelles. This knowledge would only come in the later part of the 1900s when electron microscopes were becoming widely available, allowing magnification as high as 10,000x. (The discovery of DNA took more than just higher magnification. A type of imaging called x-ray diffraction was necessary, as well as some mathematical analysis.) If you'd like to see more illustrations of protozoans from the 1800s, do a Google search using keywords: "protozoa, Haeckel."

ACTIVITY 2.1   **Watch some videos of cilates in action!**

Go to the Protozoa playlist on YouTube.com/TheBasementWorkshop and watch all the videos that are listed for Chapter 2. You will get to see most of the ciliates mentioned in this chapter. There is even a video of a Didinium devouring a Paramecium!

ACTIVITY 2.2   **Comparative anatomy**

Can you match the ciliate part with its equivalent human part? Draw a line between the matches. If you don't find what you think is an exact match, just choose the best answer available.

Ciliate parts:
peduncle
cilia
dorsal side
buccal cavity
myoneme
contractile vacuole
food vacuole
Müller bodies
rostrum

Human parts:
mouth
inner ear
muscle
kidneys
fingers
nose
foot
stomach
back

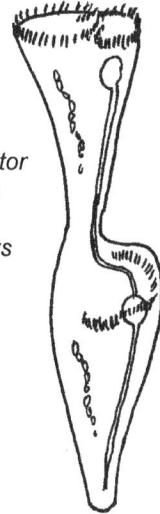

This is a Stentor going through binary fission. Ciliates always divide end to end, like this.

ACTIVITY 2.3   **"Who am I?"**

Guess which protozoan is giving you each clue. You can use the same answer more than once.

1) If I am cut up into dozens of pieces, I can grow each piece into a whole new body. _____

2) I can contract my body to 1/4 its normal size in only 1/200 of a second. _____

3) Paramecia are my favorite snack. _____

4) I am a good swimmer, but I can also "walk" along surfaces. _____

5) I am very tiny and jump about like crazy. _____

6) I bounce up and down on my very thin peduncle, which can look like a spring. _____

7) I am cup-shaped and colonial. _____

8) I know which way is "up." _____

9) I have a proboscis. _____

10) I have a rostrum. _____

11) I have two star-shaped contractile vacuoles, and one large macronucleus. _____

12) I often take in green algae cells which then live happily inside my body for a long time. _____

13) All my contractile vacuoles are lined up along my dorsal side. _____

14) I am a favorite of researchers. Many discoveries about cells were made with me. _____

15) I am long and skinny and do not have any trichocysts. _____

Possible answers: Paramecium, Dileptus, Didinium, Loxodes, Spirostomum, Stentor, Vorticella, Epistylis, Euplotes, Tetrahymena, Halteria

ACTIVITY 2.4   ZEUS versus JUPITER, round 2

Here they are, back at it again. Looks like Zeus has the longer list this time. Fill in the blanks below with the English words that contain these Greek or Latin words. In this round, you will also fill in a few of the meanings of the Greek and Latin words. (Also, we tossed in a few extra words.)

GREEK

boskein (to _____) _____

di (two or twice) _____, _____

dinos (whirling) _____

epi (upon) _____

eu (_____) _____

halteres (jumping weights) _____

hymen (membrane) _____

kolpos (gulf) _____

leptus (thin) _____

plos (swim/swimmer) _____

pro (____) _____, _____

*speira (coil) _____

stoma (_____) _____, _____

stulos (pillar) _____

tetra (_____) _____

toxon (bow that shoots arrows) _____ (pg 16)

LATIN

bucca (mouth or cheek) _____

carn (_____) _____

dorsum (_____) _____

pedis (_____) _____

rostrum (_____) _____

sessilis (sitting, to be sat on) _____

*spira (coil) _____

vorare (to _____) _____

vortex (vortex) _____

Greek "halteres" jumping weight (carved from stone)

* Notice that you can trace the word "spiral" back to both Latin and Greek.

* *

Review: Can you remember the meanings of these Greek words?

1) zoion _____   4) kystis _____   7) mitos _____

2) skopos _____   5) tricho _____   8) kytos _____

3) soma _____   6) con _____   9) protos _____

Possible answers: first, with, hair, animal, to watch, body, bag, thread, container

CHAPTER THREE: AUTOTROPHS

Ciliates love to eat. The basic "rule of thumb" is that they will eat anything smaller than themselves. Didinium is a notable exception to this rule, as it engulfs Paramecia which are equal in size to itself. Euplotes tends to be a little picky, but even Euplotes will eat a fairly wide range of prey, especially if food is scarce.

The ciliates act very much like little animals. They are not really true animals because they are made of only one cell, but they function as if they were tiny animals. Some ciliates are predators (like tigers), some eat small particles (like baleen whales do), some eat whatever they can find (like bears or raccoons do), and some eat the microscopic equivalent of plants (grazing like rabbits or cows). We often use special words to describe these eating behaviors: carnivore (meat eater), herbivore (plant eater), omnivore (eats anything), and detrivore (eats decaying stuff).

Notice that all of these "eaters" rely on sources of food outside of themselves, whether it be plants or animals. Plants, of course, do not need to eat. They get eaten, but they do not eat. Plants are the basis of all food webs (food chains). Without grass for rabbits to eat, wolves and owls would go hungry.

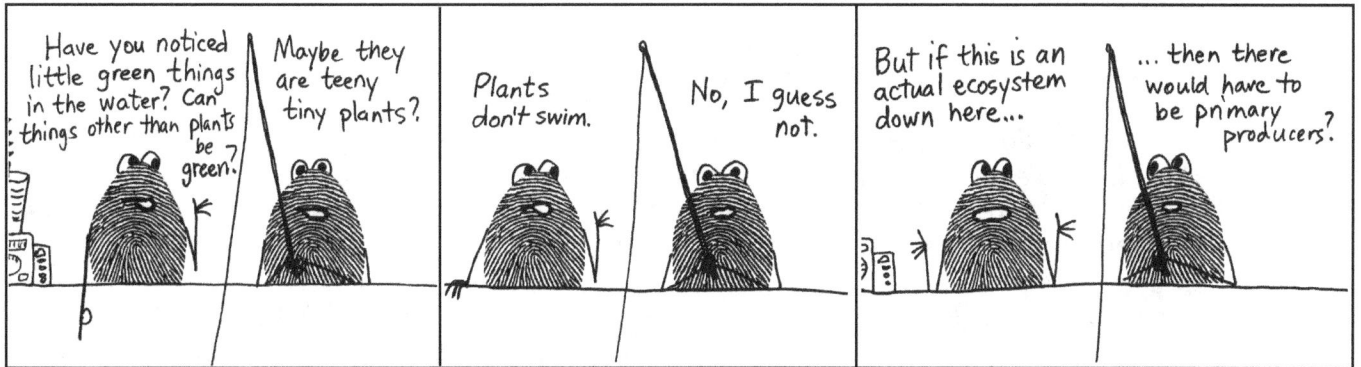

How small do plants get? Can they be microscopic? We met a very tiny plant, perhaps the smallest plant in the world, back in the introduction. Our friends were surrounded by what looked like green icebergs. These things turned out to be duckweed leaves, measuring only a few millimeters long. That is not much bigger than the head of a pin. Duckweed holds the world record for the smallest flower. It's so small you can hardly see it! The very largest forms of protozoa are smaller than a duckweed leaf. So if the duckweed is the smallest plant in the world and it is still too large for protozoa to eat, then protozoa can't possibly eat plants. (They do eat tiny bits of dead, decaying plants, but not whole, living plants.)

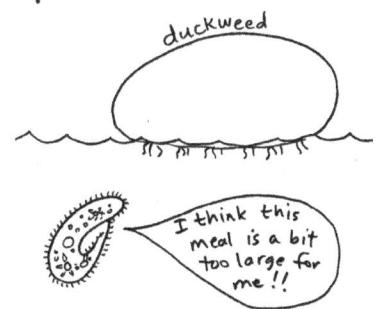

The role of plants in our ecosystems is to turn the energy of the sun into food molecules. Plants have special organelles called **chloroplasts** which have the ability to turn light (along with water and carbon dioxide) into food. The plant's goal in making this food is primarily to feed itself. Because plants can make their own food, they are called **autotrophs**. The Greek word "auto" means "self," and "trophe" means "food" or "nourishment." They nourish (feed) themselves. Autotrophs are self-sufficient as long as they have the ingredients they need (usually light, water, and carbon dioxide). They don't need to eat.

Obviously, animals and humans can't do this. Organisms that can't make their own food are called **heterotrophs**. "Hetero" is Greek for "different" or "apart from self." The heterotrophs have to get their nourishment from sources outside of themselves. Heterotrophs eat autotrophs either directly, like rabbits nibbling grass, or indirectly, like an owl that eats the rabbit that ate the grass. Without autotrophs like plants, even meat-eaters would starve.

Another term that is often used to describe autotrophs is **primary producers**. All plants are primary producers. Animals that eat plants, such as rabbits, are called **primary consumers**. Larger animals, such as owls and wolves, that eat the smaller animals that ate the plants are called **secondary consumers**. (Which organisms in our pond are secondary consumers?) Secondary consumers can be carnivores, omnivores or scavengers. Omnivores eat a bit of everything, so they act like both primary and secondary consumers.

STAY OUT OF OUR POND!

A PROTIST PROTEST

Don't confuse the words "protist" and "protest."

The primary producers (autotrophs) of the microscopic world are not plants because, as we said, plants are too large. However, our microscopic primary producers act a lot like plants, so they are often called plant-like protozoa or, more correctly, **plant-like protists**. (The suffix "-zoa" means "life," but particularly <u>animal</u> life. Our plant-like producers don't act like animals, so it's better to use the more general word for single-celled creatures: **protist**. The word protist covers all single-celled life forms except for bacteria and bacteria-like organisms.) Here in our pond, the primary producers are **algae.** Algae come in many different forms, but all of them can use the process of photosynthesis to turn sunlight into food, in the form of a simple sugar called glucose. Since they need sunlight, you will always find them at or near the top of a body of water.

The most common way algae is classified is by color: green, blue-green, yellow-green, golden-brown, brown, and red. Of course, scientists have created complicated names for these categories, like chrysophyceae, phaeophyceae, xanthophyceae, and chlorarachniophytes. Since these words are Greek for "green algae," "yellow algae," etc., we'll just use the English words.

Green algae is green for the same reason that plants are green. Both have organelles called **chloroplasts** that contain **chlorophyll**. "Chloro" is Greek for "green." Chlorophyll is the molecule that is able to "catch" light from the sun and use its energy. Plant cells and algae cells have many other cell parts in common, too. Let's compare plant cells and algae cells:

BASIC PLANT CELL

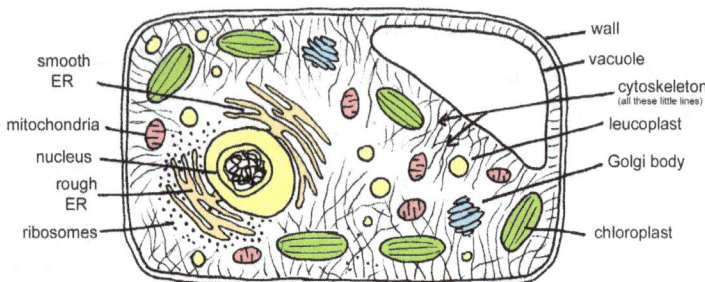

BASIC ALGAE CELL

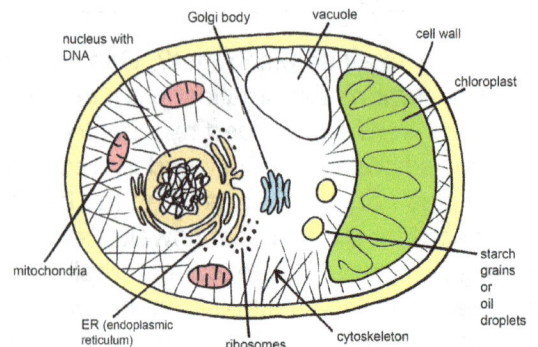

First, note the word "basic." Most algae we'll meet won't be this simple. Plant cells have a lot of variation, too. However, over-simplifying things can be helpful when we're just beginning to learn something. Second, be advised that Golgi bodies are not blue and mitochondria are not red. Cell parts don't usually have any color. (The added colors help the diagrams to be less boring.) Chloroplasts, however, are an exception. They are indeed green.

Are there any cell parts you recognize from looking at ciliates? Are you surprised to see how similar plant and animal cells are? All animal, plant, and protist cells have nuclei with DNA, ER (endoplasmic reticulum), Golgi bodies, mitochondria, ribosomes, and cytoskeleton. The biggest difference between plant and animal cells is the chloroplasts. The only reason algae are not considered to be plants is because they are **unicellular**, made of only one cell. To be in the plant kingdom you have to be **multicellular**, made of multiple cells. Algae cells sometimes get together to form colonies, but they never develop true plant parts such as roots, stems and leaves.

Let's meet our first type of green algae, Our friends will need to use their net.

They're pretty small— did you remember we've got a special filter for the net?

Yes, amazingly enough, I did.

Wow, are they tiny! They look like little green peas.

It's a good thing Mr. Pusillus gave us magnifiers. We really need them.

I think mine has two tails. Is that possible?

The smallest and simplest member of the green algae family is called **Chlorella**. It can be found just about anywhere—ponds, rivers, wet rocks, shady sidewalks, tree trunks, in your fish tank, and, as we see here... in a pet dove's outdoor drinking bowl. Okay, so you knew this chapter had to get gross eventually. It's about algae, after all. Pardon the feathers and hay in the bowl; doves are messy. The green scum that grows inside outdoor water dishes and birdbaths is likely to be Chlorella. (Probably other green protists, as well.) Chlorella has a reputation for being tough and can exist in almost any climate. It's a survivor. It reproduces very quickly, dividing every 12 hours.

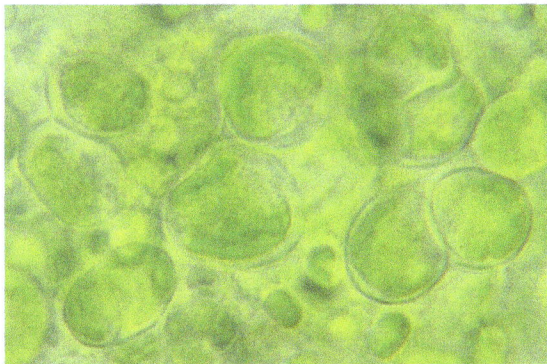

Chlorella cells are tiny, only 5 to 10 microns in diameter. (Remember, a Paramecium is about 200 microns long.) Their body structure is very much like the diagram of our basic algae cell on the previous page. Chlorella is very simple. However, Chlorella is amazing in its own way. A few decades ago, it was discovered that Chlorella is very nutritious. It has lots of protein (more protein per ounce than steak!), a high concentration of vitamins A, C, D, E and B vitamins, minerals of all kinds, omega-3 fatty acids, fiber, and lots of other things you've probably not heard about but are good for you. Some nutritionists call Chlorella a "super food."

In the 1950s, there was great concern about the growing population in both the U.S. and the world. There had been a "baby boom" after World War II, and many people believed that it would be impossible to grow enough food for this expanding population. It was at this time that Chlorella was proposed as the solution to world hunger. It could supply perfect nutrition, provided you processed it correctly so that the tough cell wall was broken apart allowing the contents to spill out. As things turned out, the agricultural industry figured out ways to make farming more efficient (fertilizers, better insecticides, genetic modifications, etc.), so our ability to grow food increased along with the population. This was fortunate because no one had discovered a way to make Chlorella truly appetizing. It's hard to disguise algae. (Alage pancakes, anyone?) People who eat Chlorella nowadays (and people do take it as a supplement) usually swallow it in pill form or drink it in green fruit smoothies.

Chlorella is a beautiful name, don't you think? If I ever have a daughter, I think I'll name her Chlorella.

Reminds me of Cinderella.

Back to the question about two tails—yes, it is possible, and, in fact, very likely. A common algae called **Chlamydomonas** *(CLAM-i-do-MO-nas)* has two **flagella**. Flagella are very similar to cilia in structure. They are both made of the same stuff that cytoskeleton is made of. (Look back at the cell drawings on page 30 and notice the little background lines labeled cytoskeleton.) Cytoskeleton is a network of protein cables that cross the cell. The end of each cable is anchored to the cell membrane just inside the cell wall. The cables are made of a special protein, designed to do this job. Many of these proteins are like long, skinny tubes, so they are called **microtubules**. Without cytoskeleton fibers, a cell would not be able to keep its shape. It would be like a plastic bag full of water, all floppy and saggy. The cytoskeleton allows the cell to "bounce back" and maintain its shape.

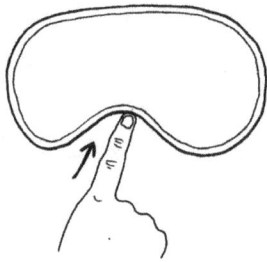

A CELL **WITHOUT** CYTOSKELETON

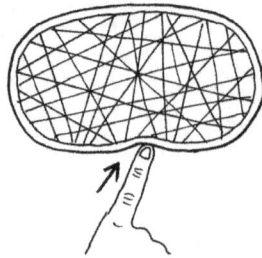

A CELL **WITH** CYTOSKELETON

Both flagella and cilia are made of these microtubules. Their structure is identical. The main difference between cilia and flagella is size and motion. Cilia are much shorter and they beat in a back-and-forth way. Flagella are very long and usually spin around at the base, a bit like a propeller.

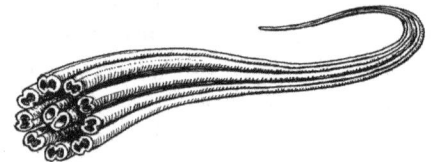

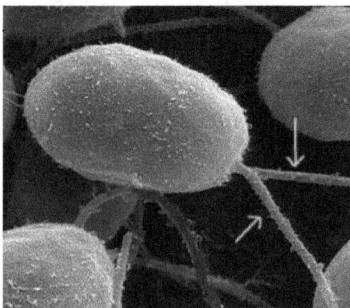

On the left is a photograph (or "micrograph") of a Chlamydomonas. You can see the two flagella coming out the bottom side. Scientists have cut off the flagella (don't worry—they grew back!) and looked at them under an electron microscope. The picture on the right shows what the end of the cut-off flagella looked like. The black circles are the ends of microtubules.

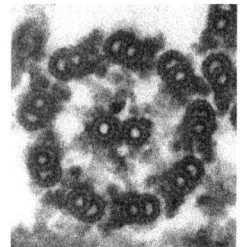

Chlamydomonas at 10,000 times normal size. This type of image can't show you the inside.

Chlamydomonas uses its flagella to propel itself through the water. Where does it want to go? Somewhere that has the perfect amount of sunlight so that it can do photosynthesis in its chloroplasts. It's like a tiny plant cell that can swim. But the world of algae gets even weirder...

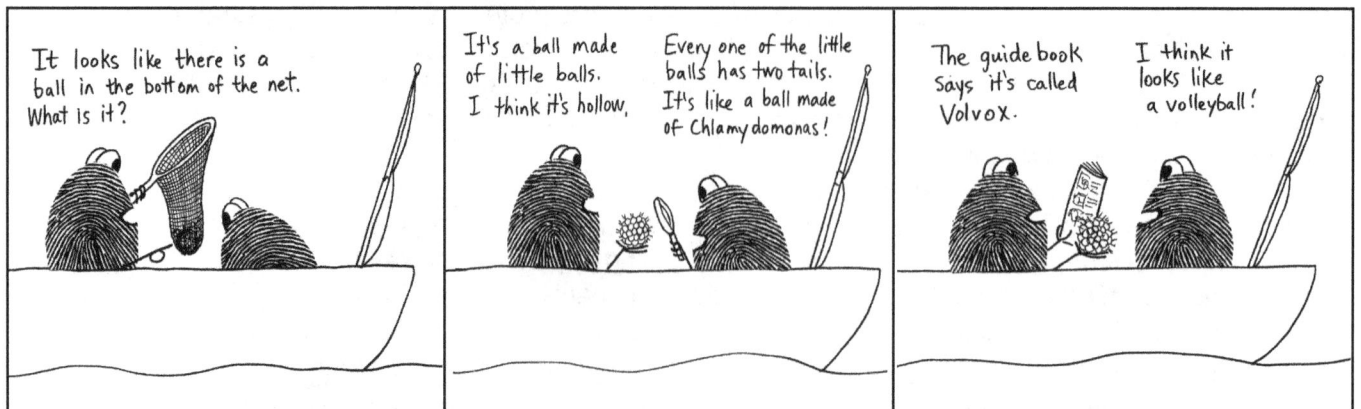

It looks like there is a ball in the bottom of the net. What is it?

It's a ball made of little balls. I think it's hollow.

Every one of the little balls has two tails. It's like a ball made of Chlamydomonas!

The guide book says it's called Volvox.

I think it looks like a volleyball!

Our adventurers have discovered **Volvox**, a **colonial** algae. In human society, colonies are groups of people living together. In microbiology, colonies are groups of microorganisms living together. The Volvox is group of 50 to 50,000 cells that look a lot like Chlamydomonas. The cells are spherical and have two flagella. They join together to make a hollow ball, with their flagella all pointing outwards. In joining together, they actually form bonds so that cytoplasm can stream back and forth between all the cells. This lets them communicate (using biochemicals).

Volvox was first seen in the year 1700 by Antoni van Leeuwenhoek *(LAY-ven-hook)*, a Dutch scientist who is often given credit as being the "father of microscopy." He did not actually invent the microscope, but he perfected the design for the hand-held model shown at the bottom of page 8, and he was the first person in history to see many microorganisms, including bacteria.

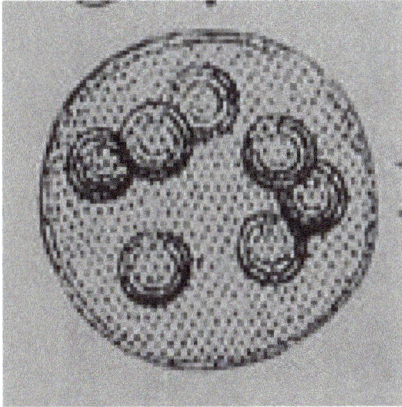

Leeuwenhoek's drawing of Volvox.

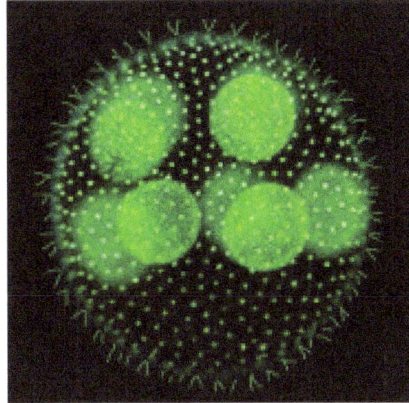

A heavily tweaked micrograph.

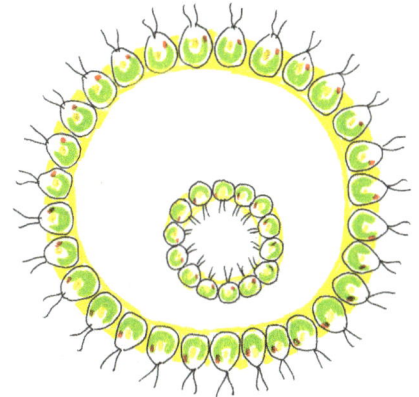

An illustration showing the structure.

Volvox cells arrange themselves into a hollow globe, using a gelatin-like substance to stick themselves together. Then they somehow decide which end is up and which is down. The top side is called the "anterior" and the bottom is the "posterior." The cells cooperate when they beat their flagella so that the ball can move towards sunlight and away from things that might harm it. The cells can sense sunlight using those little red dots.

Look at this close-up view of a Volvox cell (on right). We could use this same drawing to represent Chlamydomonas, too. (Remember, real cells aren't colorful.) Chloroplasts are green but that is about it. However, colors make diagrams so much nicer to look at. In this picture, the one large chloroplast is two shades of green. The darker green represents the areas that actually do the work of photosynthesis. The red dot is a light sensor (sometimes called an "eye spot" but it's not an eye). The nucleus and ER are in tan, the mitochondria are sort of pinkish, the Golgi body is orange, and the cell wall is gray. There are two small contractile vacuoles (shown in blue) that help pump out extra water. We saw contractile vacuoles in Paramecium and Stentor. The extra bubbles are storage vacuoles of some kind.

The purple thing is called a **pyrenoid.** (It's not really purple.) It helps to collect carbon atoms for photosynthesis. It might also help to make granules of starch (shown surrounding it, in tan). A starch molecule is a long string of sugar molecules. It's the way sugar gets stored in a cell. Before refrigeration and canning, people used to store things like beans and peppers on long strings, hanging from the rafters on the ceiling. In green protist cells, the chloroplasts make sugar (glucose) molecules and then put them into long strings for storage. On days that the cell is unable to do photosynthesis, it can "eat" some of its stored starch. (Plants store starch, too, and we often eat it. Wheat, corn and rice contain a lot of starch.)

When the time comes to make new Volvoxes, the colony must decide on one of two types of reproduction: **sexual** or **asexual**. Sexual reproduction means making male and female cells that will join together to form a little baby cell called a **zygote**. The zygote will then grow into a new adult organism. Male cells, even in algae and plants, are called **sperm**, and female cells are called **eggs**. Often, sperm have flagella so they can move. Oddly enough, in Volvox, entire colonies become either male or female. In male colonies, the cells start making sperm and in female colonies a few of the cells turn into egg cells. The sperm swim over and find the egg cells and they join to make a zygote. Zygotes are tough little cells. They can survive cold and drought, whereas regular colonies can't. When the Volvox senses that winter is coming, it starts doing sexual reproduction.

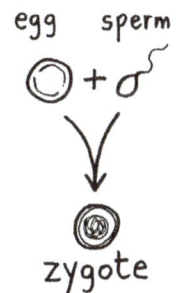

egg   sperm

zygote

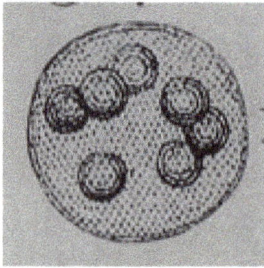

Asexual reproduction means <u>without</u> male and female cells. ("A" means "without.") To reproduce asexually, a Volvox colony makes those little balls you see inside the large ball. The little balls are called **daughter colonies**. Sorry, no sons, just daughters! Of course, they aren't really daughters; they are neutral, neither male nor female. When cells divide to make new cells, the resulting cells are always called daughters. It's just a tradition, like ships being referred to as "she."

Yes, it is possible, because the microscopic world has some really bizarre creatures. What they are seeing is Euglena *(yu-GLEEN-ah)*, a flagellated green protist. Euglena is very well known. In fact, it would probably make the "top three" list if you asked a biologist to name three kinds of protists. Paramecium would also make the top three list, as well as another protist we have not discussed yet. So Euglena is pretty famous.

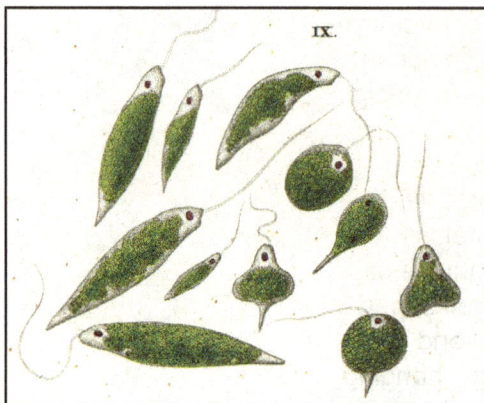

A drawing from the 1800s. (Ehrenberg)

The name Euglena comes from the Greek word "eu" *(yu)*, meaning "good," and the Greek word "glene" meaning eyeball. Good eyeball? Not really. Euglena has what is commonly called an "eye spot." In the 1800s, they assumed this red spot was a real eye. We now know that it is just a patch of red pigment that acts as a light filter. The actual light sensor is under the red spot. The sensor can only tell what direction light is coming from. It can't see shapes or colors or movement. It's not a proper eye at all. However, it's just what the Euglena needs. Since it has chloroplasts and likes to do photosynthesis, it wants to be in a place with plenty of light. A light sensor is a big help.

Euglena has a flagella, though it is so long and thin that early scientists missed seeing it. It comes out of the anterior end, from an indented place that functions a bit like a mouth. So it's like having a tail (or a propeller) coming out your mouth. If Euglena has chloroplasts and can do photosynthesis, why does it have a mouth? If conditions are not good for photosynthesis, Euglena can take in small bits of food and digest them, just like a Paramecium can. So here is an organism that is <u>both</u> an autotroph and a heterotroph. There aren't many of these on the planet. Euglena is pretty special!

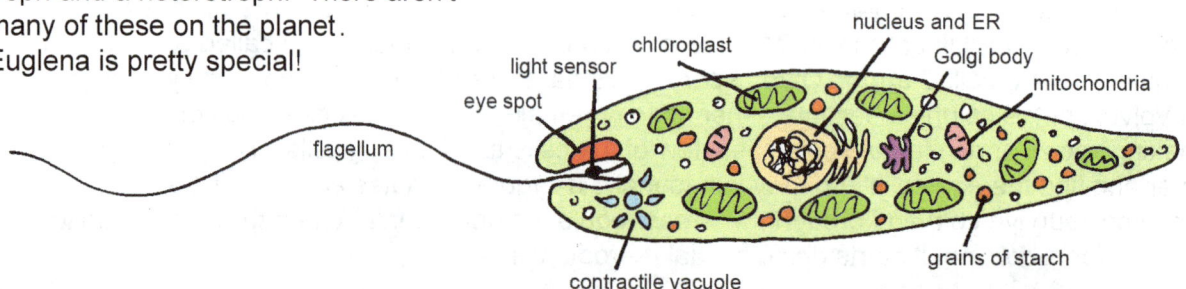

If Euglena can be heterotrophic, then can it be classified as a green algae? It can do photosynthesis and is unicellular, just like all the other green algae. As we've seen, algae can have flagella. Some algae, such as Volvox, have red eye spots. Euglena is so similar to green algae that, in fact, some science books will classify it as such. However, Euglena has a feature that sets it apart from green algae: *it does not have a cell wall*. The outer layer of Euglena is a flexible pellicle, more like what you find in ciliates. Euglena is different enough from green algae that it gets to be in its own group, called the Euglenoids.

Euglena's outer pellicle has a very interesting pattern. This picture was taken with a scanning electron microscope, or SEM. The SEM can only show you outer surfaces, not interiors, so despite the fact that Euglena are transparent, this SEM image shows only the surface texture. In this SEM, it looks like someone wound a string around and around and around the Euglena's body. These strings are made of those microtubules we learned about on page 32 (plus a few other proteins). They are able to slide around without falling off. This sliding action allows the Euglena to bend and stretch and change its shape. (You can see how flexible Euglena are by watching the Euglena videos on the YouTube playlist.)

Just one bit of trivia before we move on... Recently, scientists have discovered that Euglena is an even better organism than Chlorella to use as an emergency food source. The problem with Chlorella is the thick cell wall. Breaking that cell wall involves an expensive procedure. Euglena doesn't have a cell wall, so it is cheaper to process. In 2005, a Japanese company called Euglena Co. began producing Euglena-based food products. They've also discovered that Euglena produces oil droplets that can be collected and refined into fuel for jet engines.

Panel 1: "I think we're getting farther away from the shore..." "Should I toss this guy back in?"

Panel 2: "Why didn't you throw it back in?" "I don't know. I guess he looks fun to play with."

Panel 3: "He? How do you know it's a male colony?" "Well, there aren't any daughter colonies inside and none of the cells look like they are turning into eggs."

Panel 4: "Sounds like a reasonable theory." "Catch!"

Panel 5: "It's Volvox volleyball!"

Panel 6: "Oh, drat! I missed!"

(You can play your own version of Volvox Volleyball. See the activity section!)

There's a whole category of green algae that are *filamentous*. The word filament comes from the Greek word "filum" meaning "thread." Filaments are like long threads. The algae they found certainly does look like long threads. Can a cell be this long and skinny? Not in the algal world. These strings are made of many cells attached end to end. A close up view of Spirogyra looks like this:

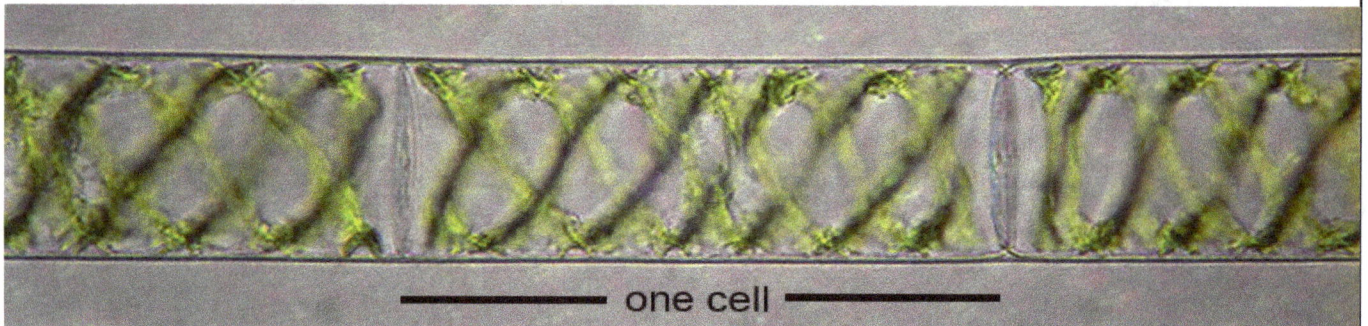

━━━━━ one cell ━━━━━

It's beautiful, don't you think? The irony is that large colonies of Spirogyra (and other filamentous algae) are what make "pond scum." Yes, that icky stuff on the surface of ponds is actually beautiful Spirogyra! It's not disgusting at all when you look at it under the microscope. And it doesn't smell. If your pond scum smells, this is due to the presence of bacteria, not algae.

As you might guess, the green things are chloroplasts, even though they don't look anything like the chloroplasts we've seen so far. These chloroplasts are long and skinny and are coiled up to make a spiral shape. The spiral shape is where Spirogyra gets the "spiro" part of its name. The "gyro" part means "turn." The chloroplasts look like they are twisting and turning. (But they aren't; they are sitting still.)

The little dots all along the cholorplast are called *pyrenoids*. We saw a pyrenoid inside Chlamydomonas. The pyrenoids help to collect carbon atoms from the water so they can be used in photosynthesis. The pyrenoids may also help to store glucose sugars as grains of starch. The stored starch can be used by the cell on days where there is reduced sunlight and therefore reduced photosynthesis.

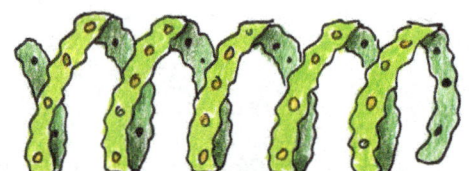

The nucleus of Spirogyra isn't visible in the photograph on the opposite page. It is transparent and therefore very hard to see. In this drawing, the nucleus has been colored yellow so you can see it. (The chloroplast has also been removed.) The blob around the nucleus is cytoplasm, the gel-like goop that usually fills cells. The central cytoplasm has "arms" extending out to the edge of the cell. These "arms" keep the nucleus "floating" in the center. The space around becomes a very strangely shaped vacuole filled with water. What a weird cell!

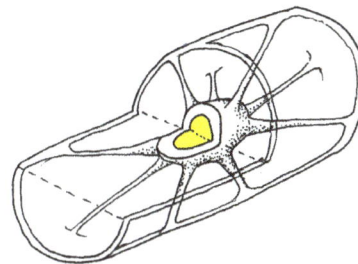

Spirogyra's reproductive cycle is a bit more normal — for an algae. Like other algae, it can reproduce sexually or asexually. Remember, sexual reproduction just means that there is DNA being traded back and forth. Trading DNA increases the variety of information an organism has. If you collect trading cards of some kind, or participate in a role-playing game that involves cards, you understand the value of trading. If you've got a lot of similar cards it limits what you can do. Usually, you want to trade so that you have a large variety. Living organisms are the same way with their DNA. DNA has information that helps them to survive things like diseases or harsh conditions. Swapping DNA increases their chances of survival.

To reproduce asexually (no DNA trading) a cell simply splits in half. The cells divide end to end, making those long strings our friends saw in the water. A cell can reproduce itself in a matter of hours if conditions are right. This is the primary mode of reproduction for Spirogyra. The cells divide and divide and divide and divide. If a strand of Spirogyra gets cut into pieces, each piece will start growing into a new long strand. When a piece (fragment) of an organism can grow into a whole new organism, this is called *fragmentation*.

The Spirogyra can also reproduce sexually (combining the DNA of two cells). Two strands line up so their cells are side by side. Then little tubes start to grow out from each cell. When they touch, they fuse to form a connecting tube (1). Then one cell (we don't know how they decide which one) will empty its contents into the other one (2). This is called *conjugation*. "Con" is Latin for "with," and "jugo" is Latin for "join." Conjugation means "joining with."

Now we have an empty cell, and a cell with two of everything (3). The cell with double everything will change to become a *zygote* (3). A zygote is what you get when two reproductive cells join together. In most animals and plants, the cells that join to form a zygote are an egg and a sperm. However, Spirogyra doesn't make eggs or sperm (like Volvox does). After the cells join, Spirogyra's zygote turns into a *spore* called a *zygospore* (4). Spores (of any species) have durable outer coatings and are able to survive harsh conditions such as very hot, very cold, or very dry. The Spirogyra zygospores often sink to the bottom of the pond and wait until spring to start growing. This process of going through conjugation to form zygospores is commonly used not only by filamentous green algae but by other protists, as well.

Another filamentous green algae with a cool name (that is closely related to Spirogyra) is *Zygnema* (zig-NEE-ma). Zygnema has star-shaped chloroplasts. There are hundreds of other species of filamentous green algae, but these two types are the most famous ones. Some of the other types produce branches and are harder to identify as filamentous algae.

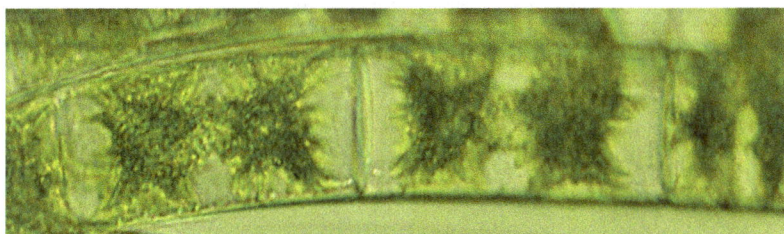

Okay, so if I can't name my band Spirogyra, I'll name it Zygnema. That's just as cool.

Sorry, Zygnema is taken, too. It's a heavy metal band from India.

Before we leave the fascinating topic of green algae, there are two organisms that we absolutely must mention. One of them looks very much like a filamentous green algae, but isn't, and the other doesn't look like algae, but is.

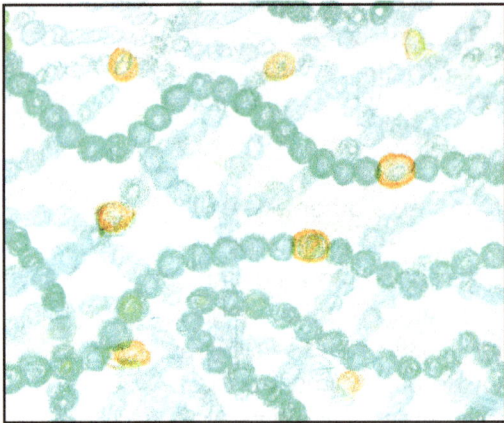

This type of cyanobacteria is called Nostoc. It can be found not only in fresh water ponds, but also in soil. The tan circles are special cells where nitrogen is processed into a form that plants and algae can use. The bluish-green circles are cells that do photosynthesis.

In your bucket of pond scum you might find something that looks very much like filamentous green algae except that the cells are more round, looking a bit like a pearl necklace. These are often called blue-green algae, but they are really a type of bacteria called *cyanobacteria*. The Greek word "cyan" means "blue." The reason that they have been confused with algae is because they do photosynthesis.

Most bacteria don't do photosynthesis, so at first it was assumed they must be a type of algae. However, when electron microscopes got powerful enough that scientists could look inside these blue-green cells, they discovered that there were no chloroplasts. Not only that, there was no nucleus and no organelles such as Golgi bodies and mitochondria. These were not *eukaryotic* (yu-care-ee-ot-ik) cells. Animals, plants, fungi and protists all have eukaryotic cells with a true nucleus and organelles like mitochondria, ER and Golgi bodies. "Eu" means "good" (though in this case the meaning is more like "true"), and "karyo" means "kernel" or "nut," referring to the nucleus.

If a cell isn't eukaryotic, it must be *prokaryotic* (pro-care-ee-ot-ik). "Pro" means "before," referring to the belief that prokaryotic cells existed before eukaryotic cells. Prokaryotic cells don't have a true nucleus. They have DNA, but it isn't inside a nucleus; it just floats around as a tangled blob. Since prokaryotic cells don't have organelles, either, they seem much simpler than eukaryotes. Blue-green algae cells were found to have no true nucleus so they had to be prokaryotic bacteria cells, not eukaryotic algae. Since they are not really algae, let's call these cells by their proper name: cyanobacteria.

This picture shows how simple a cyanobacteria is on the inside. The only organelles it has are ribosomes. Ribosomes are the little factories that make proteins.

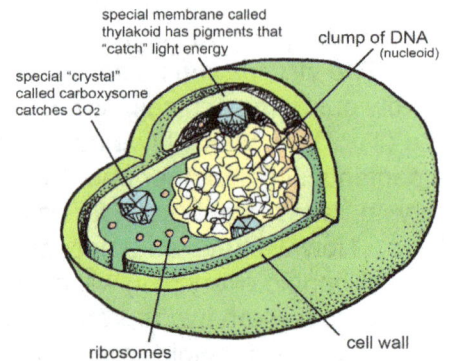

A basic cyanobacteria cell. Notice the lack of organelles.

Cyanobacteria are Olympians when it comes to photosynthesis. They crank out tons of oxygen that ends up being released into the air, replenishing the atmosphere. The story of how they manage to do photosynthesis without all the usual cellular equipment that plant and algae cells have is a long one, and we can't go into it here. However, they do have a bluish-green pigment that is similar to chlorophyll and a few other important molecules for photosynthesis. Somehow they manage to get the job done!

Cyanobacteria are also important because they have the ability to collect nitrogen from their environment and put it into molecules that plants (or algae) can use. Plants and algae need nitrogen in order to make protein and DNA. Even though there is plenty of nitrogen in the air or water all around them, plants and algae can't use it. Cyanobacteria (and other "nitrogen fixing" bacteria) can grab nitrogen atoms and put them into a form that plants and algae can use.

Some plants, such as beans and peas, have this type of bacteria living in their roots all the time. Lucky them! In the old days, farmers would rotate their crops so that each field got a chance to grow beans or peas which would replenish the nitrogen in the soil. Most farmers nowadays just put nitrogen-rich fertilizers onto their fields. Very recently, another option has become available: "bio-fertilizer." Bio-fertilizers are alive! They contain living bacteria, such as cyanobacteria, that will hopefully become a permanent part of the soil, replenishing the nitrogen continually (saving the farmer a lot of time and money). [Azospirillum by Criyagen® shown here.]

A bio-fertilizer

What doesn't look like algae, but is? All types of seaweed. Since this pond adventure takes place in a fresh water pond and seaweed is mostly found in ocean (marine) environments, we'll make this very brief.

This a topic that makes many students want to yell, "No fair! No fair!" You've learned that algae is unicellular (made of single cells). We've seen that sometimes the single cells can live together in a colony, such as Volvox and Spirogyra. In both these cases, all the cells are pretty much the same. When you look at some types of seaweed, however, you see cells specializing and acting like roots, stems and leaves. How can the giant kelp, shown in this picture, be in the protist kingdom?

To understand why the kelp is classified as a **brown algae** and not a plant, you must have a microscope and a lot of knowledge about plants. When you look at the microscopic structure of these "leaves," they aren't like plant leaves at all. (We must call them "blades" instead of leaves.) Kelp's "roots" don't take in water and nutrients like a plant's roots do. (We must call them "holdfasts" instead of roots.) Kelp doesn't make flowers or seeds, and it doesn't have proper stems with xylem and phloem tubes. However...

If you've studied botany you might be thinking, "But moss doesn't have true roots and leaves, and it doesn't have xylem and phloem tubes, and yet it is classified as a plant. Kelp is very similar to moss. Why can't kelp be a moss-like plant?" Good point. Yes, kelp has a lot of similarities to moss.

How can these not be leaves?!

Even the reproductive cycle is similar. We're "splitting hairs" here, exploring some very technical definitions. The best explanation you are likely to find is that the cells in all types of algae must nourish themselves, whereas plants have some kind of distribution system for sharing water and nutrients (with moss as the exception). Also, algae rely on fragmentation as a primary reproductive method, whereas plants do not. (Although it is true that some plant leaves, such as those of the Jade plant, are capable of growing into a new plant.)

This brown seaweed is called **sargassum**. Those round things that look like berries are actually filled with air. They act like flotation devices, keeping the leaves (oops — "blades") at the surface of the ocean where there is sunlight for photosynthesis. Air-filled "bladders" are common in brown algae.

As we end our discussion of seaweed, it is good to bear in mind that the classification system is not set in stone. It is not unusual for scientists to rethink the classification of an organism and switch it to a new category. Perhaps the classification of seaweed and algae will change in your lifetime. (And the kelp will go on being kelp no matter what we think about it!)

ACTIVITY 3.1   ZEUS versus JUPITER, round 3

What? Zeus gets the longer list again? This must be rigged. Wait till Juno hears about this. Let's hope she doesn't send her sons Mars and Vulcan to get revenge. Especially Vulcan, since he contols volcanoes!

As usual, write the meaning of the word root inside the parentheses, then write a word that contains that root.

GREEK

auto (_____) _____

trophe (nourishment) _____

hetero (_____) _____

chloro (_____) _____

eu (good/well/true) _____, _____

glene (_____) _____

gyro (_____) _____

pro (_____) _____

zygo (yoke) _____ ⟶

LATIN

con (   ) _____

filum (_____) _____

jugo (_____) _____

pro (_____) _____

Cows wearing a yoke.

Don't confuse the word YOKE, which is a wooden bar used to keep two cows together, with the word YOLK, which is the yellow part of an egg. Notice the difference in spelling.

ACTIVITY 3.2   A FEW "THINKING" QUESTIONS

1) Why is the Greek word for "yoke" appropriate for naming the zygote? _____
_____

2) Which protist in this chapter is both a primary producer and a primary consumer? _____

3) Chlorella does not have an oral groove like a Paramecium does. Why not? _____
_____

4) If a pond is classified as "eutrophic" what might that mean? _____

5) Make a food chain by putting these words into the blanks: Didinium, Stentor, Paramecium, Chlorella

_____ is eaten by _____ which is eaten by _____ which is eaten by _____

ACTIVITY 3.3   MS. EUGLENA'S "EU" CHALLENGE

Ms. Euglena has a puzzle for you. Can you match these "eu" words with their meaning? Figure out as many as you can on your own before you ask for help Do the ones you are most certain of first, saving the mysteries for last. Think about what word stems such as "photo," "phonic," "logos," "thermic," or "rhythm," might mean. What other words contain these stems? When you get stumped, you can ask Google what a word stem means, or ask an adult for a hint.

1) ___ Eubacteria
2) ___ Eugene
3) ___ Euglena
4) ___ Eulogy
5) ___ Euphemism
6) ___ Euphonic
7) ___ Euphotic
8) ___ Euplotes
9) ___ Eucalyptus
10) ___ Eurythmics
11) ___ Euthanasia
12) ___ Euthermic

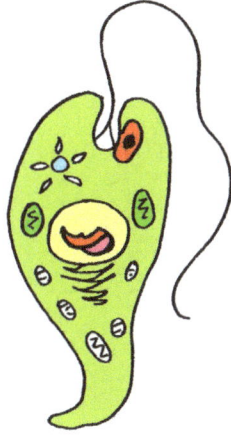

A) sounds pleasant
B) not too hot, not too cold
C) good eye
D) true bacteria
E) saying good words about someone
F) well bred (good genetics)
G) good swimmer
H) a good name for something that isn't so good
I) ending an animal's life for its own good
J) a good amount of light
K) well covered (referring to flower buds)
L) music education using body movements

ACTIVITY 3.4   VIDEOS

Check out the Chapter 3 videos on green and brown algae on the Protozoa playlist at YouTube.com/TheBasementWorkshop.

ACTIVITY 3.5   "I'M SEEING PROTISTS EVERYWHERE!"

Have you been studying protists so much that ordinary objects are starting to remind you of protists? Just for fun, what protist does each of these objects remind you of?

yam

eggplant

yellow squash

ribbon

soccer ball

necklace

paper fastener

slipper

lollipop

ice cream cone

celery

green peas

string of tickets

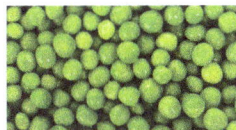

ACTIVITY 3.6:   REVIEW PUZZLE

If you have trouble figuring out these clues, the number in parentheses is the page where you can find the answer.

DOWN:

1) These organelles produce energy for the cell. (10)
5) This process is how Paramecia split in half. (11)
6) This long, thin protein is a bit like a muscle. It is what allows Vorticella to contract quickly. (23)
10) This protist has tiny dots called Müller bodies that help it sense gravity and know which way is up. (20)
11) Protists that lives together in groups are c___. (26)
12) An organism that eats producers is a ____. (30)
13) Seaweed is not a plant, it is a ____. (39)
14) This green algae was proposed as a solution to world hunger after World War II. (31)
15) There are 1000 of these in a millimeter. (5)
16) This means "not moving." (21)
17) The watery gel that fills a cell. (9)
20) This word means "bottom." (7)
22) This is Didinium's favorite meal. (16)
26) This word means "top." (7)

ACROSS:

2) A tiny sac that contains a sharp "harpoon." (12)
3) This organelle is like a packaging warehouse. (10)
4) This word means "for feeding" and refers to a long, skinny nose-like appendage. (17)
7) An organism that can't make its own food. (34)
8) This organelle is a tiny version of the bigger one, and is used only during reproduction. (11)
9) This is the outer "skin" of protists like Paramecium. (9)
12) An organism that eats meat (flesh). (15)
18) A tiny protist that moves with sudden jumps. (24)
19) A protist that looks like it is having a bad hair day, but is a "good swimmer." (24)
21) Means "joining with." It's how protists trade DNA. (37)
22) This type of cell does not have a true nucleus. (38)
23) This ancient Greek had a very loud voice. (21)
24) This type of organism can make its own food. (29)
25) Chlamydomonas has two of these (tail-like). (32)
27) Organelle where photosynthesis takes place. (29)
28) When an egg and sperm join, they form a ___. (33)
29) This type of vacuole expels extra water. (9)
30) These little organelles make proteins. (10)

CHAPTER FOUR: DIATOMS, DESMIDS, and DINOFLAGELLATES

There's an old proverb that says, "People who live in glass houses shouldn't throw stones." What the proverb means, of course, is that people who are vulnerable to criticism shouldn't go around criticizing others. No one would really live in a glass house. However, in the world of protists, there are creatures that DO live in glass houses! They are called **diatoms** and they are one of the most numerous organisms on the planet. Diatoms live in every pond, stream, river, lake and ocean of the world.

Panel 1: Hey- does it look to you like we are getting farther from shore? / Hmm....

Panel 2: The grass used to look like it filled half the sky. Now I can actually see over the tops of the grasses. / Yeah, I guess you're right. There's more open sky now.

Panel 3: And look at the water. It looks more clear. Less pond scum. / You mean "fewer filamentous green protists"

Panel 4: In fact, I see less of everything. / Yeah, the water looks pretty clear.

Panel 5: Do you suppose we've seen everything there is to see? / Maybe we should check the guidebook.

Panel 6: There's a whole section here that we haven't seen yet. / I just saw something big and slimy. But it's gone now.

Panel 7: They're called diatoms and they make shells out of silicon dioxide, which is basically glass. You can find them in open waters. / If they are clear like glass, maybe that's why we can't see them.

Panel 8: Well, I'm going to swish the net around and see if there is anything here.

Panel 9: Wow! There are little things in here! But I don't think they are alive. They look like geometric shapes. Like jewelry or something.

Indeed, diatoms look very much like little glass boxes. They have finely detailed geometric features. If they were a thousand times larger, we would clean them out and sell them as novelty items in gift shops. On the next page, you will see a poster of diatoms drawn back in the 1800s.

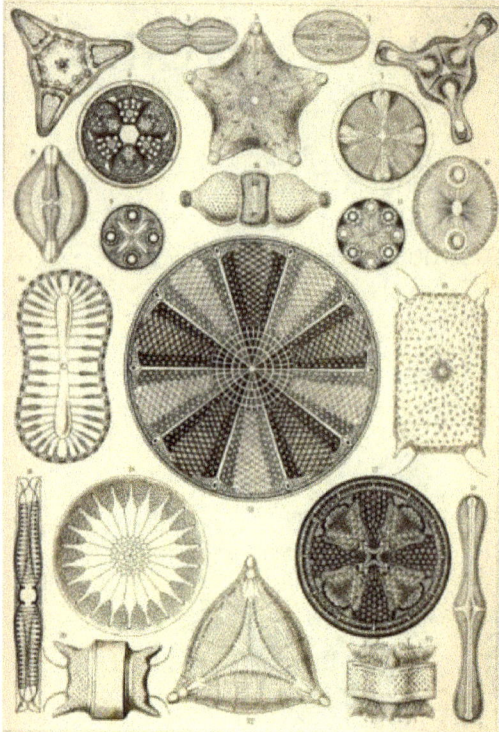

As our friends said, it's hard to believe these things are alive. Yet each one is a living cell and contains most of the organelles we saw inside the ciliates and the algae.

Diatoms are photosynthetic. Just like plants, they use sunlight to make their own food. But they aren't green. So far we've been able to use color as a really good clue as to whether or not an organism has chloroplasts and does photosynthesis. Diatoms fool us. Most are brown or yellow-brown. In fact, another name for diatoms is the "yellow-brown algae." Their yellows and browns come from extra molecules that help chlorophyll collect sunlight. Diatoms do have some chlorophyll, but not enough to make them look green (although some ocean diatoms can look bluish-green). These other molecules, called "accessory pigments," are far more abundant so they are the ones we see. Plants that have red or yellow leaves have lots of accessory pigments. They do still have green chlorophyll in their chloroplasts, but the bright red and yellow pigments drown out the greens. Tree leaves that turn colors in the fall do this because the green chlorophyll starts disappearing, letting you see the reds and yellows that were there all along.

We know that plants play a very important role in the environment as they take in carbon dioxide and produce oxygen. For animals, carbon dioxide (CO_2) is a waste product. If there was no way to recycle carbon dioxide, we'd be in big trouble. Plants do play an important role, yes, but diatoms and other protists play an even larger role. Most of our atmospheric oxygen comes from the protist kingdom. Diatoms alone produce at least 25% of the earth's oxygen. Green algae and cyanobacteria add another 40 to 50%. All the grass and trees in the world produce only a fraction of our oxygen. (Three cheers for the protist kingdom!)

I'm makin' O_2!

Let's take a look at the anatomy of a diatom. As we've already said, it has an outer shell made of SiO_2, silicon dioxide, the same stuff that sand and glass are made of. The diatom has the ability to take the mineral silicon out of the water and use it to build its house. (Clams and snails have a similar ability do to this, but with the mineral calcium.) The glass shell is called a *frustule* and is made of two halves called *valves*. The valves fit together like the lid and the bottom of dish. Diatoms come in some weird shapes, but let's take a look at a very simple one.

Imagine a dish with a lid that comes down over the sides. If you've ever seen a petri dish (used for bacteria experiments) this shape is very similar.

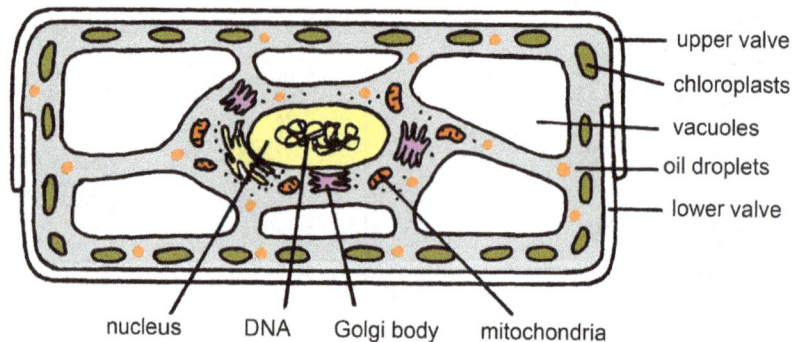

upper valve
chloroplasts
vacuoles
oil droplets
lower valve

nucleus   DNA   Golgi body   mitochondria

Diatoms have a nucleus with the usual surrounding endoplasmic reticula, plus Golgi bodies, mitochondria, chloroplasts, and big empty spaces called vacuoles. You will notice that they also have oil droplets. Diatoms prefer to store their food as oil, not sugars. This is handy because oil happens to be less dense than water so it floats in water. If you are an organism that lives in water but would like to stay at the surface, you need all the buoyancy you can get. The oill droplets help the diatoms to float. The vacuoles would also help with buoyancy.

The oil is also a very rich source of calories. If you want to eat a meal that will keep you going a long time, make it high in fat. Oils and fats can store more energy than sugars. Diatoms are important producers in both freshwater and marine ecosystems. In the cold ocean waters around Antarctica, tiny lobster-like crustaceans called krill eat diatoms as their primary food. (A krill might eat as many as 100,000 diatoms a day!) Krill are then eaten by whales.

Diatoms that live in the ocean (marine diatoms) tend to be circular or triangular in shape. They are called **centric diatoms**. The poster on the opposite page shows mostly centric diatoms. Freshwater diatoms tend to be longer and thinner and are called **pennate diatoms**. If you look at a drop of pond water under a microscope you will almost certainly see diatoms that look similar to these:

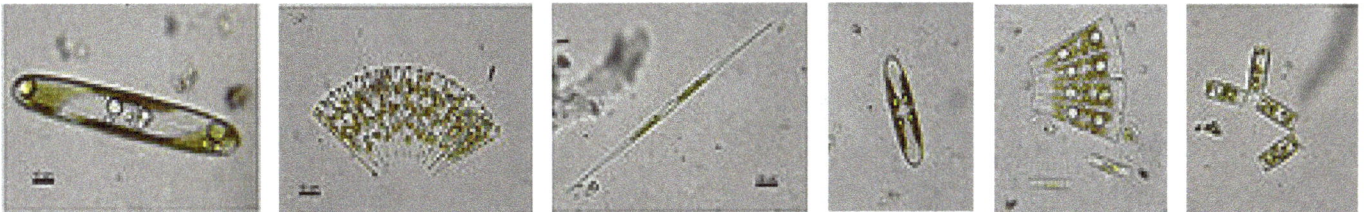

These are all pennate diatoms. They have scientific names, of course. You don't need to learn their names, but for those of you who like words and names, the long oval shapes are Navicula, the fan-shaped ones are Meridon, and the skinny one is Synedra.

Sometimes diatoms live as colonies, like Tabellaria, shown on the left, and Fragilaria, on the right. As the diatoms duplicate themselves, the chain grows longer. If cells break off, they grow into new chains.

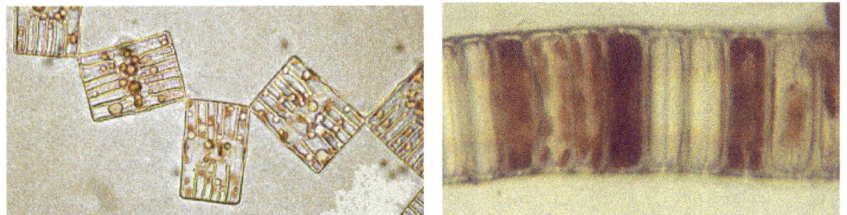

As with other protists, diatoms use asexual reproduction most of the time. The top and bottom valves are held together by connective tissue that can allow them to separate. The organelles all duplicate themselves and migrate towards either the top or the bottom. Then the two halves separate and each half grows a new valve. The lid grows a new bottom, and the bottom turns into a lid and then grows a new bottom. The original bottom was smaller than its lid, so when the bottom turns into a lid, this new diatom is going to be smaller than the original. Imagine this happening over and over again, with the bottom always turning into a lid. One of the duplicates always end up smaller than the original. We are going to end up with some very small diatoms. In fact, they will be too small to be able to hold the nucleus and all the organelles. Now what can be done? Sexual reproduction to the rescue! When the diatoms get too small, they know it is time to swap DNA with another diatom and make gametes (eggs and sperm). The egg and sperm cells will join together and make a brand new diatom that can grow to maximum size. Then the shrinking starts again...

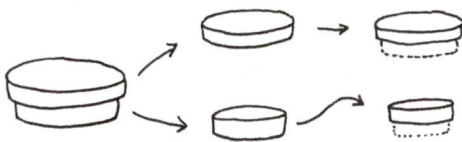

Pennate diatoms can do something that centric ones cannot. See all those holes in the frustule? Those are the diatom's connection with the outside world. No organism could live in a tightly sealed glass container, not even a protist. All diatoms have holes, but pennates also have a slit down the middle called a **raphe**. In a manner not yet fully understood by scientists, these diatoms can push some of their cytoplasm out through the slit and use the cytoplasm to propel themselves. The pennates can glide along through the water at a pace similar to walking.

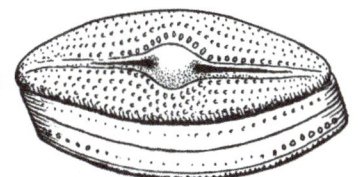

Panel 1:
- Do you think we'll be able to take these back?
- You mean when we go back to our real size? Will they get big?

Panel 2:
- Yeah. What would happen if we kept a pile of them in the boat with us?
- The stuff we brought down, like the net— they shrank...

Panel 3:
- Would everything in the boat get big again?
- We could get rich selling these on ebay...

Panel 4:
- How many do we have here anyway?
- Hey— look at this one!

Panel 5:
- It looks like a green banana with bubbles inside.
- Pennate diatoms aren't supposed to be green, are they?

Panel 6:
- Looks like it's a "desmid." Desmids are actually a type of green algae. They're not diatoms.

Desmids are very easy to confuse with diatoms. They have symmetric shapes and they are about the same size—from 50 to 200 microns. They do look like they might be diatoms. However, their outer covering is not made of glass and they don't make oil droplets so they can't be diatoms. Though they may not look like it, desmids are actually very similar to Spirogyra. This makes them a green algae. (So they're in the wrong chapter, aren't they? Sorry.)

Here is another poster by Ernest Haeckel, who lived in the late 1800s. Can you find our green banana? Its name is Closterium. The thing in the middle is the nucleus. The strings of dots are **pyrenoids**, which help the chloroplasts get carbon atoms from the CO_2 in the water. Carbon is necessary for making glucose sugar. Diatoms and other kinds of alge can have pyrenoids, too.

This is a photograph of Cosmarium, another very common desmid. Those circles are its pyrenoids.

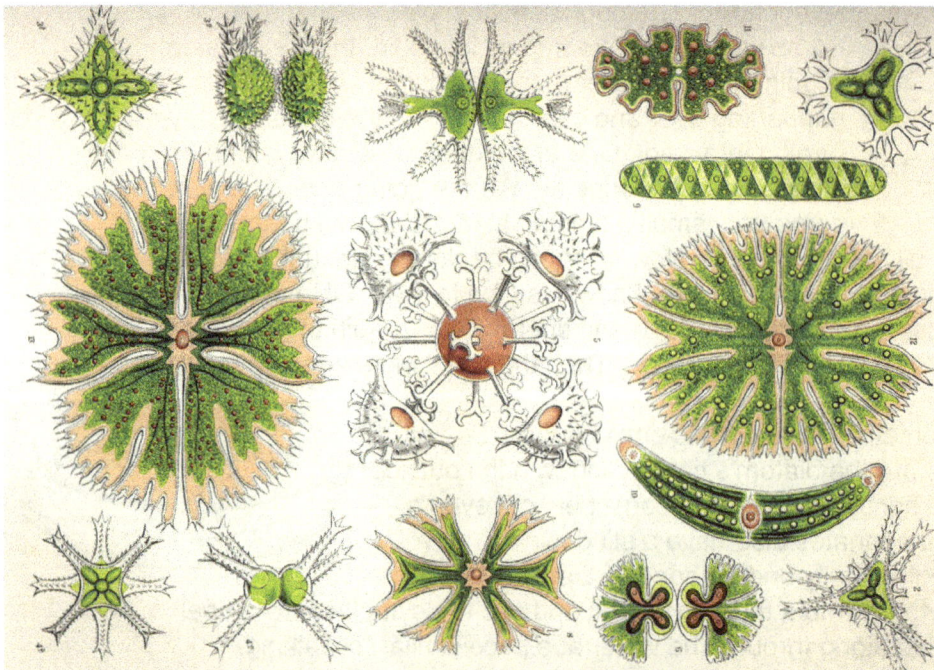

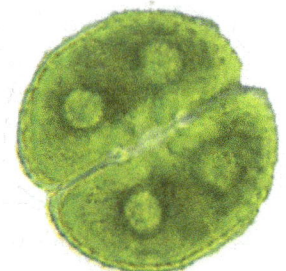

Panel 1: So I guess it would be pointless to keep these desmid things if they don't have a glass shell.

Panel 2: Yeah, I think it would dry up like a green bean. And... what's that thing you pulled up? A mouse?

Panel 3: It's like an armored mouse with no legs and an extra tail.

We've discovered another protist that begins with the letter D! This is a ***dinoflagellate***. Unlike the word "dinosaur," which means "terrifying lizard." the word "dinoflagellate" does not mean "terrifying flagellate." The word "dinos" is actually Greek for "whirling." (The "dino" in "dinosaur" comes from "deinos," meaning "terrifying.") Dinoflagellates have two flagella that propel them around in such a manner that they look like they are whirling and spinning. Well, most dinoflagellates. Or maybe at least some of them.

There are so many different kinds that you can hardly say one thing that is true about all of them. Some are autotrophs, some are heterotrophs and some do both. Some are covered with armored plates, some are not. Some have spikes, some do not. Some are triangular, some are spheres. The characteristic they all seem to have in common is that they have two flagella and a certain type of outer wall. What they are best known for is their bizarre shapes.

The cell wall of most dinoflagellates is not made of glass like diatoms, nor cellulose like desmids and algae. It's not a pellicle like a paramecium or euglena. The wall is made of something hard to describe. Imagine taking a bunch of vacuoles (empty vacuoles, with no food in them), squashing them flat, and then drying them out a bit. Put these into layers and stick them together. Basically, they are complicated. All our friends need to know is that they are not made of glass, so it won't do them any good to take them home. They won't make beautiful glass dishes.

This poster shows some of the shapes that had been cataloged by the end of the 1800s. We've discovered many more since then. Scientists estimate that there are about 2,000 species of dinoflagellates. Most live in the ocean, but a few do live in fresh water.

Some dinoflagellates do something really nasty. They make toxic chemicals that attack the nervous system of many animals, including humans, causing severe illness or even death. No one has been able to figure out why they do this. Usually when organisms make toxins it is to help them catch and kill prey.   But this doesn't seem to be what is going on here. At any rate, as long as there aren't too many of them in the water, they aren't a serious threat. Eating a few dinoflagellates isn't going to cause a problem.

Another poster by Haeckel

However, they have a second nasty trait—their population can explode practically overnight. When conditions are right, they can reproduce unbelievably fast. Soon the water is saturated

with them. Some species can make the water look red. When this happens, it is called a **red tide**. (The "red" part is okay, but the "tide" part is a bit misleading because it doesn't really have anything to do with the tide going in and out. The correct name for this phenomenon is **harmful algal bloom**. Some blooms are green or brown, not red.) The state of Florida has red tides off its southwest coast almost every summer. In 2004, many dolphins died. Baleen whales and loggerhead turtles have washed up on the shore.

While the red tide is occurring, fishermen have to be careful not to catch any fish or other sea life from that area. Fish and crabs don't understand that the red stuff is polluting their food supply and they go on eating as they normally do. Many of them don't die right away and the toxins begin to build up in their bodies. If they are caught and sent to a restaurant or grocery, the humans consuming them will get a high dose of toxins and begin having neurological symptoms. This is called **neurotoxic shellfish poisoning**.

Here is the little villain who causes the red tides of Florida: *Karenia brevis*. This image was taken by an SEM (scanning electron microscope). Remember, this type of microscope can only show you the outside shape and texture, nothing else. Can you find its two flagella? The one around its middle is easy to see. It also has one coming out the top, but it is folded down in this picture. We'll need to take a look at a diagram that is more clear.

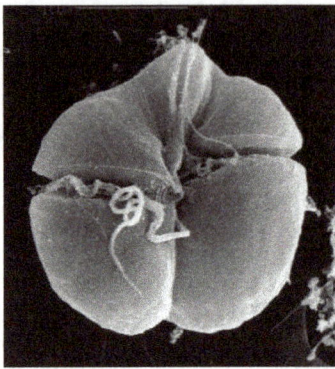

Let's take an up-close look at the armored mouse with no legs and an extra tail. (Or a turtle with two shells and two tails?) The flagella around the middle is the one it uses primarily to power itself forward. The one sticking out the bottom seems to be used mostly for steering.

Look at the diagram of the inside. (Remember, those colors aren't real; they are just to make the diagram look nicer.) How many things does the dinoflagellate have in common with a diatom or a paramecium? We see oil droplets, a Golgi body, a chloroplast, some mitochondria, and something called a "pusule." The

pusules seem to function a bit like contractile vacuoles, helping to regulate how much water is inside. However, they seem to be different enough from contractile vacuoles that scientists think they need a separate name. We also see something labeled "dinokaryon." This is the nucleus, but notice how the DNA is stuck to the walls. No other creature has its chromosomes stuck to the insides of the nucleus. So, of course, they had to think up a new name for the nucleus, calling it the "dinokaryon." We saw the root word "karyo" in the last chapter when we learned about prokaryotes and eukaryotes. It means "kernel or nut," sort of a nickname for the nucleus. Lastly, we see those flattened, dried-out vesicles at the edges that make the armored plates.

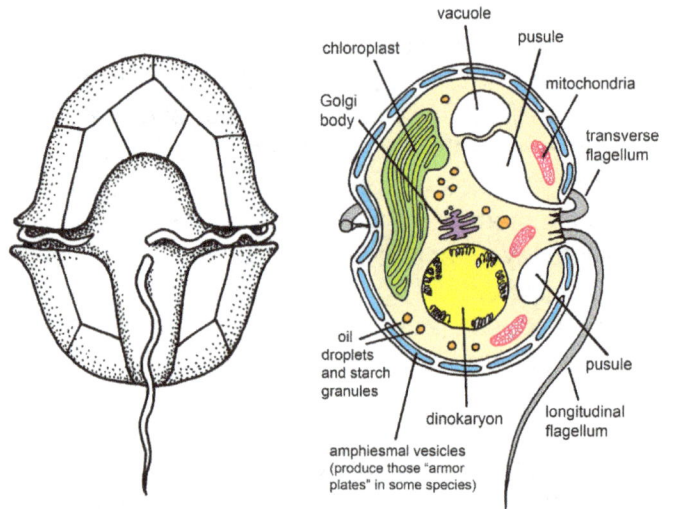

And now... we can end with something really fun. Some dinoflagellates have a spectacular talent—they **glow in the dark**! Marine dinoflagellates exhibit "bioluminescence" when disturbed by a passing boat or fish. They produce light in a manner similar to fireflies. So red tides can be fun at night. You can see this phenomenon for yourself by watching the YouTube playlist.

ACTIVITY 4.1   Thrilling videos of diatoms and dinoflagellates!

Once again, go to the Protozoa playlist on YouTube.com/TheBasementWorkshop. You will see gliding diatoms and glowing dinoflagellates.

A satellite map showing the green algal blooms along the coastlines of Lake Erie.

Two of the videos on the playlist are about algal blooms in Lake Erie (Pennsylvania, Ohio and Ontario) and Lake Champlain (New York, Vermont and Quebec). The University of Ohio put together a very nice 7-minute video explaining the history of Lake Erie's blooms. They interview researchers and farmers who are working to solve the problems (and generally succeeding). The big issue seems to be too many nutrients in the water, primarily phosphates and nitrates from farms. The farmers need to add these nutrients to the soil in order to have their crops produce enough food. However, when it rains, the rainwater washes a lot of these nutrients out of the soil and into the lakes. The nutrients feed the algae and they reproduce too quickly, creating massive blooms. The blooms disrupt life in and on the lake, causing fish to die and tourists to stay away. Lake Erie had a severe bloom problem back in the 1970s, caused by sewage run-off. They fixed the sewage problem and the lake went back to being clear and healthy. So hopefully, things will turn out well once again.

The video on Lake Champlain is longer, but is worth watching. The same issues plague this lake, threatening the health of the ecosystem and the economies of lakeside communities.

If these videos happen to be missing from the playlist, such use the search feature with the names of the lakes and the words "algal bloom."

****

Noctiluca scintillans, one of the bioluminescent dinoflagellates

This is *Noctiluca*, one of the bioluminescent dinoflagellates you will see in the videos. ("Nocti" means "night" and "luca" means "light.") Believe it or not, this spherical (not flat) dinoflagellate is actually a predator. It is large enough to catch and eat fish eggs, diatoms, other dinoflagellates, and many other types of algae.

Noctiluca ranges in size from 200 to 2000 microns. (That would be up to 2 millimeters, easily visible with the naked eye.) It eats its prey by taking them in and putting them into food vacuoles, just like paramecium does. The thing that looks like a flagella is actually a tentacle for gathering food. The real flagella is so small you can't see it; it is practically useless. So this dinoflagellate breaks our rule that dinoflagellates have two flagella. However, if we count the tentacle as a flagella-like structure, we can still permit the Noctiluca to be a dinoflagellate. Someone needs to let these difficult protists know they are messing up our classification system!

One of *Noctiluca's* favorite meals is diatoms. Often, the diatoms will survive for long time, even reproducing inside. Since *Noctiluca* is a heterotroph and does not have chloroplasts, it is helpful to develop a symbiotic relationship with a photosynthesizing protist.

Noctiluca can be found worldwide, in both warm and cold seas and oceans, and is capable of causing harmful algal blooms.

****

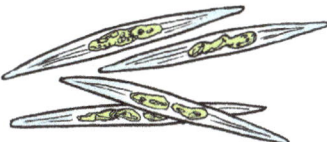

Diatoms are occasionally to blame for harmful algal blooms. The diatoms shown here are known to be harmful. Like dinoflagellates, they make a toxin that harms the nervous system, causing seizures and amnesia (loss of memory). Amnesia is easy to diagnose in humans, but a bit trickier in a sea lion or a pelican! If it is treated right away, victims make a good recovery. However, the toxin is indeed dangerous and people have died of it.

ACTIVITY 4.2   ZEUS versus JUPITER, round 4

The gods are disgruntled. This chapter was a big disappointment because it had so few new words in it. They will have to think of something else to fight over. As usual, fill in the meanings of the words and the words that contain those roots.

GREEK

dinos (_____) _____

deinos (_____) _____

saur (_____) _____

karyo (kernel or nut) _____

pyro (fire) _____
<small>We did not formally introduce this one.</small>

LATIN

luca (_____) _____

nocti (_____) _____

pustula (blister) _____

ACTIVITY 4.3   Comparing the D's

Diatoms, desmids and dinoflagellates. How are they the same and how are they different? Put a checkmark in each box that applies.

	DIATOMS	DESMIDS	DINOFLAGELLATES
1) Produces oxygen for the earth's atmosphere			
2) Cells can join together to make long colonies			
3) Has a nucleus with DNA			
4) Has an outer shell made of glass			
5) Has flagella			
6) Are green, so they belong to the green algae group			
7) Come in two basic patterns: centric and pennate			
8) Some members can move or can glide			
9) Has Golgi bodies and mitochondria			
10) Makes oil droplets			
11) Has chloroplasts and does photosynthesis			
12) Has beautiful geometric shapes			
13) Has top and bottom that can come apart			
14) Can cause harmful algal blooms			
15) Reproduces by splitting in half			
16) Has very visible pyrenoids			
17) Can be both autotrophic and heterotrophic			
18) Can glow in the dark			

CHAPTER FIVE: AMEBOIDS

We've discovered a protist called **Amoeba proteus**. Both these words means "change." It's like calling it "the changeable changer." What it changes is its shape. But before we launch into a discussion of its anatomy (or lack thereof), we need to briefly discuss spelling.

We are going to call this creature an "*ameba*." This word began its history in the 1700s as "Amiba," taken from the Greek "amoibe," meaning "change." Later, the spelling was changed to "Amoeba." During the 1800s, scientists liked to use *ligatures*—two letters stuck together, like "æ" or "œ" to emphasize the fact that these science words had come from Greek or Latin. (Notice that the Greek is "am**oi**be.") You might remember that Paramecium used to be Paramoecium. Other science words had "œ," too. Esophagus used to be œsophagus, and diarrhea used to be diarrhœa. Then someone started a campaign to get rid of the "o's." In general, it worked, at least in America, but for some unknown reason, the "o" in Amoeba stayed there, right up until the present day. Currently, some American scientists are trying again to get rid of the "o" in Amoeba. The word "Ameba" looks easier to spell and pronounce than the word "Amoeba." So now some books write "Amoeba" and others write "Ameba." Both are considered to be correct and you can use either one. We are going to choose the less scary-looking version and go with "Ameba."

æ œ

"ligatures"

As long as we are talking about spelling and grammar, we'll just go ahead and mention that if we use the term "ameboid," we don't have to use a capital "A" because we are not talking about a particular organism, just a type of organism. And, unlike Paramecium and Stentor, the word ameba is often not capitalized. (The bright side is that it's hard to be wrong!)

Ameboid protists are unlike any we've seen so far. They are in their own phylum group, called the Amoebozoa. All ameboid organisms are soft, squishy lumps that ooze around. Some ameboids are "naked" (eek!) and others are more decent and make a shell covering for themselves. The ameba clinging to the boat is a naked ameba.

The ameba was discovered back in the 1700s. This drawing was done in 1755. (What was going on in your country in 1755?) These early microbiologists noticed that their amebas did not seem to have mouths. They took in food by surrounding it with their oozy *pseudopods*. ("Pseudo" is Greek for "false," and "pod" is Greek for "foot.") They watched their amebas eat just about anything. The amebas could even catch ciliates and flagellates.

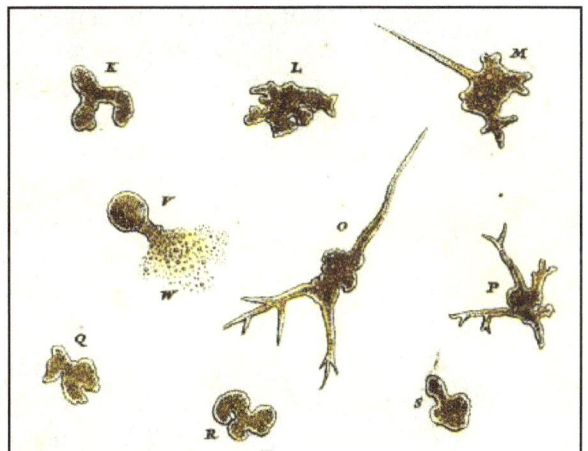

Drawing by C.G. Ehrenberg, 1755.

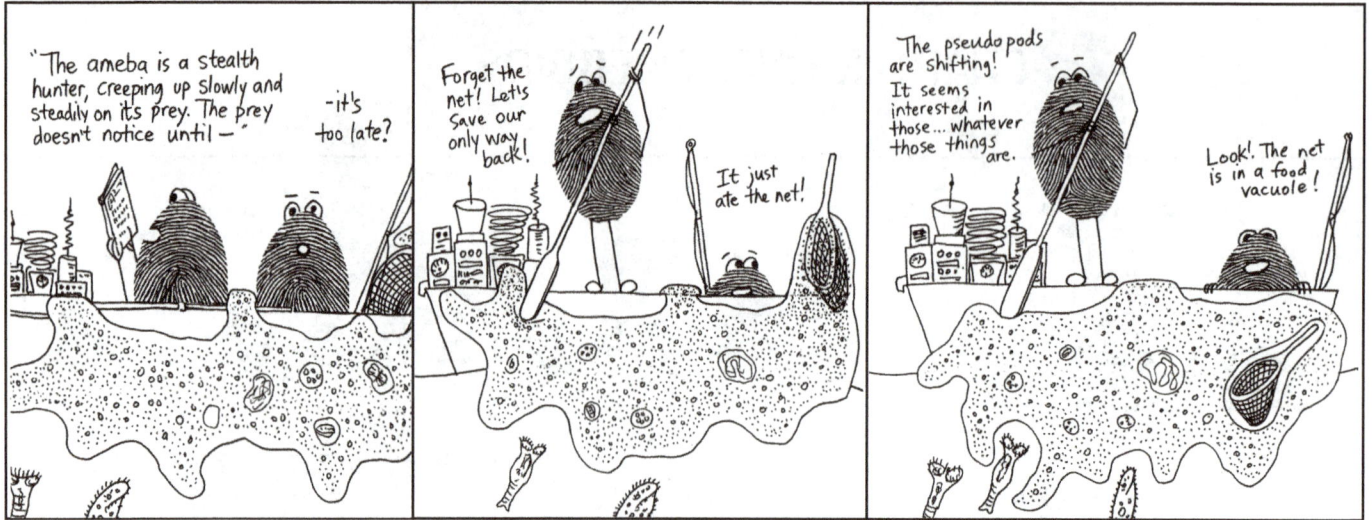

"The ameba is a stealth hunter, creeping up slowly and steadily on its prey. The prey doesn't notice until —"

—it's too late?

Forget the net! Let's save our only way back!

It just ate the net!

The pseudo pods are shifting! It seems interested in those... whatever those things are.

Look! The net is in a food vacuole!

You wouldn't think that such a slow-moving creature could capture anything fast, but somehow it captures enough of them to stay alive. (Perhaps the ameba is like tortoise in the famous story by Aesop, where the tortoise wins the race against the rabbit. The ameba is slow but steady, and persistence pays off in the end.) Some species of amebas will also eat bits of decaying plants. Cyanobacteria and single-cell algae are nutritious and easy to catch, too.

So what kind of anatomy do we have here? An ameba pretty much looks like a blob. Basically, that is exactly what it is—a blob that constantly changes shape. It has a nucleus and other cell parts, but they aren't in any particular place. Amebas do have a contractile vacuole, but it can be hard to see. What you mostly see are food vacuoles. Like most other protists, it is transparent so you can see what it has been eating.

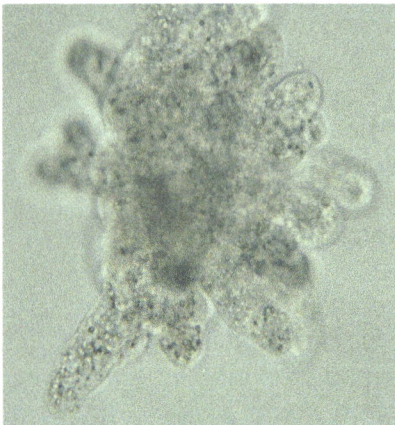

Amebas have a "granular" texture. The inside is not clear and smooth, except around the edges. (Scientists make a big deal about this clear outer edge, since it is one of the very few anatomical features an ameba has. They call it the **ectoplasm**.) Mostly it is full of blobs and dots. It's hard to explain the granular texture. Some of the blobs are probably small food particles, others would be waste particles. There would be mitochondria and Golgi bodies floating around, too. There could be storage vesicles filled with oil or starch. It's a big, goopy, mixed up mess inside, but it works for the ameba.

The pseudopods (those blobby "arms") are extended when the ameba wants to move or to capture prey. There are tiny sensors on the ameba's outer membrane. These sensors can catch chemical particles drifting by and bring them inside. If one part of the membrane is bringing in a lot of particles that "smell" like food, the ameba will move in that direction. To move, an ameba uses its **cytoskeleton**. Look back at page 32. We learned that every cell has a network of little "strings" running all through it (the cytoskeleton), which helps the cell to keep its shape. These strings can be dissolved and rebuilt very quickly. The ameba dissolves the strings in the areas that it wants to flow. Once the flowing is over, the string network is rebuilt within seconds.

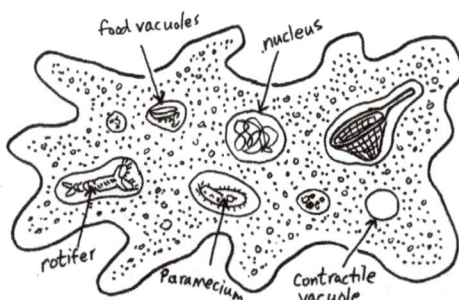

food vacuoles

nucleus

rotifer

Paramecium

contractile vacuole

What has our ameba eaten? A fishing net, for one thing, but what are those other things? You'll have no trouble recognizing the Paramecium. The other one is a **rotifer**. Rotifers are not protozoans and are not part of the protist kingdom. Though they are the size of a protozoan, they are tiny multicellular animals. Here we have a single cell that ate a multicellular animal! Rotifers range in size from 100 to 500 microns. Amebas can grow as large as 3000 microns, but are usually more like 200-500 microns. The ameba shown here is a very large one.

Even though they don't properly belong in a currculum about protists, let's sidetrack just a bit and see some micro-animals. We'll look at rotifers, gastrotrichs, and ostracods.

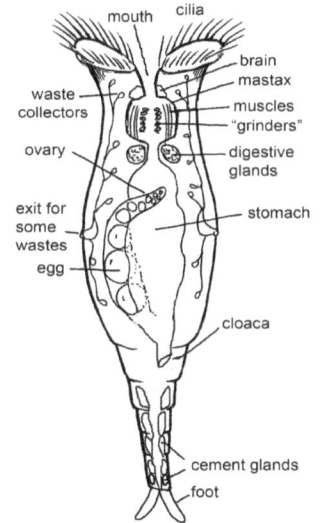

The name *rotifer* means "wheel-bearer." The cilia beats in such a way that those two circles on the top look like they are going around like wheels. (It's an optical illusion.) You don't have to become an expert on rotifer anatomy—just notice how different rotifers are from protists. Rotifers have a proper stomach, not just a food vacuole. They have a brain, although it is extremely small and simple. They have a throat called the mastax, which has teeth-like grinders for chewing food. There are salivary and digestive glands, too.

Rotifers have a simple waste collection system that might remind us a bit of our own kidneys. They have muscles, too, for moving and for chewing. Some rotifers have feet and use cement glands to stick themselves in place. This rotifer is a female with ovaries and eggs. (Males don't occur very often, but when they do they are small and don't live very long.)

Rotifers are made of hundreds of cells, even though they are the same size as large single-celled protozoans. Rotifers range from 200-500 microns and come in all kinds of bizarre shapes. Some make tubes to live in.

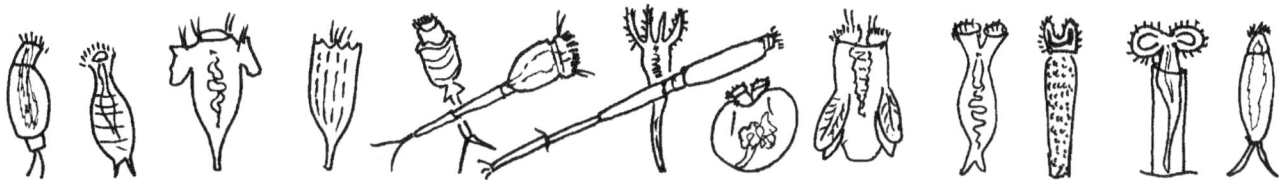

Gastrotrichs are not as well-known as rotifers. ("Gastro" is Greek for "stomach," and you already know that "tricho" means "hair.") They are smaller than rotifers, often less than 300 microns, and they like to live in stagnant (not moving) water or in sediment at the bottom of a pond. They have hair on their bodies, funny tufts of hair on their heads, and long pointed "feet." The bottom is covered with cilia that beat in unison allowing them to glide smoothly along. They have many of the same body parts as rotifers, such as stomach, waste collectors, digestive glands, and a simple brain.

Unlike rotifers, they have no males and females. They are both male *and* female at the same time. Animals that make both eggs and sperm are called *hermaphrodites*. In ancient Greek mythology, Hermaphroditus was the son of Hermes (the messenger god with wings on his feet) and Aphrodite (the goddess of love and beauty). When Hermaphroditus was 15 years old, a nymph (water goodess) fell in love with him. He did not want to have anything to do with her, so she had to trick him. One day she began hugging him and praying to the gods that they would never be parted. The gods decided to answer the nymph's request and merged the two bodies into one, making it impossible to ever part. So poor Hermaphroditus was both male and female from that time on.

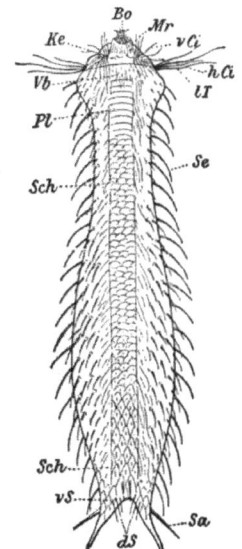

Drawing from 1911

Ostracods are microscopic *crustaceans*. "Crustacea" is the group (class) that lobsters and crabs belong to. Ostracods are sometimes called "seed shrimp" because the shape of the shell looks similar to a flat seed, such as a squash or cucumber. Their diet consists mostly of algae; they gobble down desmids and diatoms like candy. Some species will eat other things, too, such as cyanobacteria, pollen grains, and tiny ciliates.

There are tens of thousands of species of ostracods. You'll find some type of ostracod living in nearly every pond, river, lake and ocean of the world. They range in size from 200 to 1000 microns.

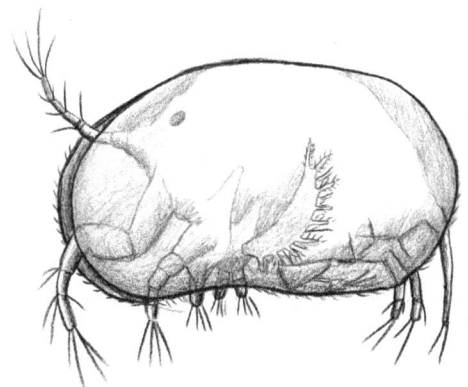

Ostracods are by far the most complicated creatures you will meet in this book. They really are little animals. They have 5 to 8 pairs of legs that they keep inside their shell most of the time. They've got antennae, a digestive tract, muscles, a tiny brain and a simple eye. They have no need of a heart or lungs, though, as they are so small they can absorb oxygen from the water.

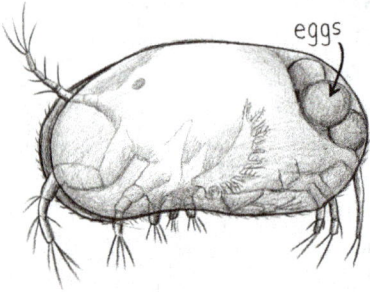

eggs

Ostracods have males and females, but females are much more common. Males are generally produced only when environmental condition start to get difficult. This is true for many microscopic animals. When the pond starts to dry up, or they sense that winter is coming, microscopic animals will begin to produce males that can fertilize eggs. These fertilized eggs will be much tougher than regular eggs, and will be able to survive harsh conditions. So wait a minute—their regular, every-day eggs are produced without any males? How can this be? If chickens lay eggs without a rooster around, they won't hatch. The hens will still lay, but the eggs won't have baby chicks inside of them. (Eggs that you buy at grocery stores were never capable of producing baby chicks. The egg farmers don't keep roosters around; they sell them to meat stores. Some people think that by eating eggs you are killing baby chicks. This is not true. If you want "fertile" eggs that will make chicks, you have to go to stores where organic, all-natural products are sold and look for eggs that are specifically labeled "fertile.")

Ostracod eggs are very different from chicken eggs. They will develop and hatch without any ostracod males around. When eggs hatch without having been fertilized by male sperm, it is called *parthenogenesis*. This word is related to the Greek word *Parthenon*, the temple built for the goddess Athena. Athena was the virgin goddess who never married. These unfertilized eggs will grow into normal adult females. They only need males when they want to make eggs that will survive cold and drought.

And now, back to protists...

Well, I don't know about you, but I think I've had just about enough adventure for one day. Maybe we should think about heading back soon?

We haven't seen the strangest things in this book. I wish we could have seen radiolarians.

We lost our net, you know.

Did you know that there are ameboids who build glass skeletons that look like fancy holiday ornaments? But we lost our net.

Hey- look at our stash of diatoms! There's hardly any left! That ameba must have eaten them. So much for getting rich on ebay...

You're right. That stupid ameba devastated our supply!

It would have been nice to catch some of these!

Even though we won't actually see them, do you still want to learn about them? Sure, why not?

Some ameboid creatures build shells to live in, just like clams and snails do. How can a single-celled creature with almost no anatomy construct a house for itself? Nature is full of deep mysteries! Hundreds, perhaps thousands, of species of ameboid protists collect tiny particles of sand, dirt or even diatoms, and glue them together to make a shell. They keep their nucleus inside the shell and just extend pseudopods for feeding. They make little cytoplasm strings on their interior portion and connect these strings to the shell so they can't be pulled out of the shell. Whatever eats them must eat shell and all. Some creatures will still eat them, but others will decide the shell is too hard and will leave them alone. Amebas like the one pictured here are simply called **shelled amebas**.

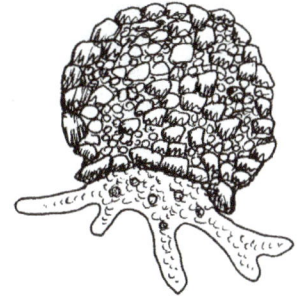

Other ameboid creatures have more complicated body and shell shapes, and they also have more complicated names. The **Heliozoans** were discovered centuries ago and were called "sun animacules" because they looked like they had rays going out from them, like sunshine rays. ("Helios" is Greek for "sun.") We now know that these rays are actually very long, skinny pseudopods (called **axopods**). In order to maintain this long, skinny shape, the axopods have stiffening rods made of cytoskeleton filaments. The heliozoans are predators, capturing small protozoans. (Notice the small protozoan stuck to one of the top axopods.) They use the axopods to grab them and move them closer to the main body. When they touch the main body, a food vacuole is formed right there at the edge; then it is brought inside. (Notice the food vacuole at the bottom edge.) As with regular amebas, the body has a granular endoplasm, and clear ectoplasm. Some heliozoans make scaly shells at the edge of their ectoplasm.

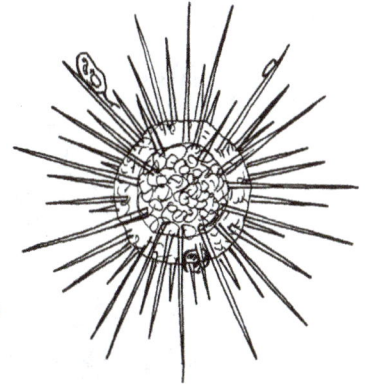

The **Foraminiferans** (for-AM-in-IF-er-ans) have a complicated name that simply means "hole bearers." "Foramin" means "hole" and "fer" means "bear." This is the word "bear" as in "to bear a burden," or "to bear a load." It is one of the oldest words in the English language. The most well-known word that uses "fer" is the name Christopher, which means "Christ bearer." An ancient legend tells of a saint who carried the Christ child across a rushing river. He became known as the "Christ bearer," or "Christopher." The root word "fer" is a good one to know; you will meet it from time to time as you study science. (Example: rotifer)

The "forams" are similar to the heliozoans, except that they build solid shells, called **tests**, made of calcium carbonate, the same stuff that clam shells are made of. They can take these minerals out of the water around them. They leave little holes in their tests (like diatoms do) so that they have contact with the outside world. Like all living things, they have to be able to get oxygen and fresh water. How do they eat? Apparently, in a similar way to the Heliozoans, though it's hard to imagine.

When forams die, their tests pile up at the bottom of the ocean or lake. At some point in earth's past, many were fossilized, becoming embedded in sedimentary rocks in various part of the world. It is said that the limestone that was used to build the pyramids of Egypt is full of fossilized forams. Some tropical beaches are made primarily of foram tests. Fossilized forams can give clues to petroleum engineers as to where to dig oil wells. Experts on microscopic fossils have correlated certain forams with rocks that are likely to have oil under them.

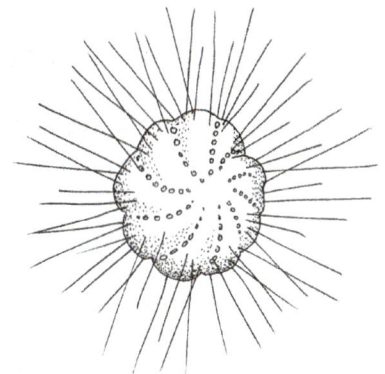

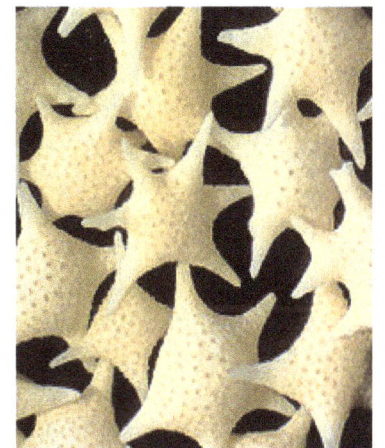

2085f Japon Hatoma_ by Psammophile
- Microphotographie personnelle
(From Wikipedia)

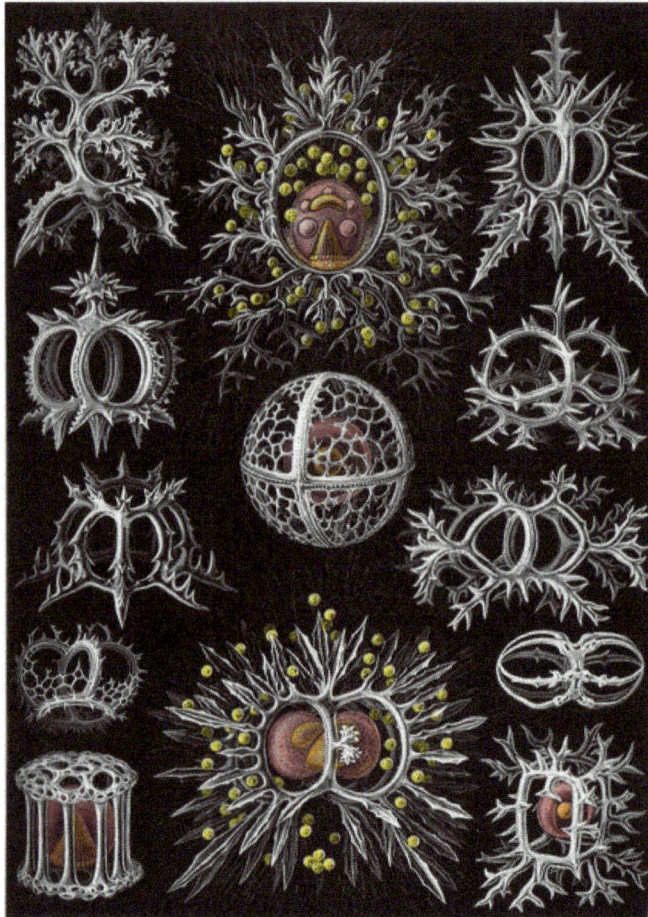

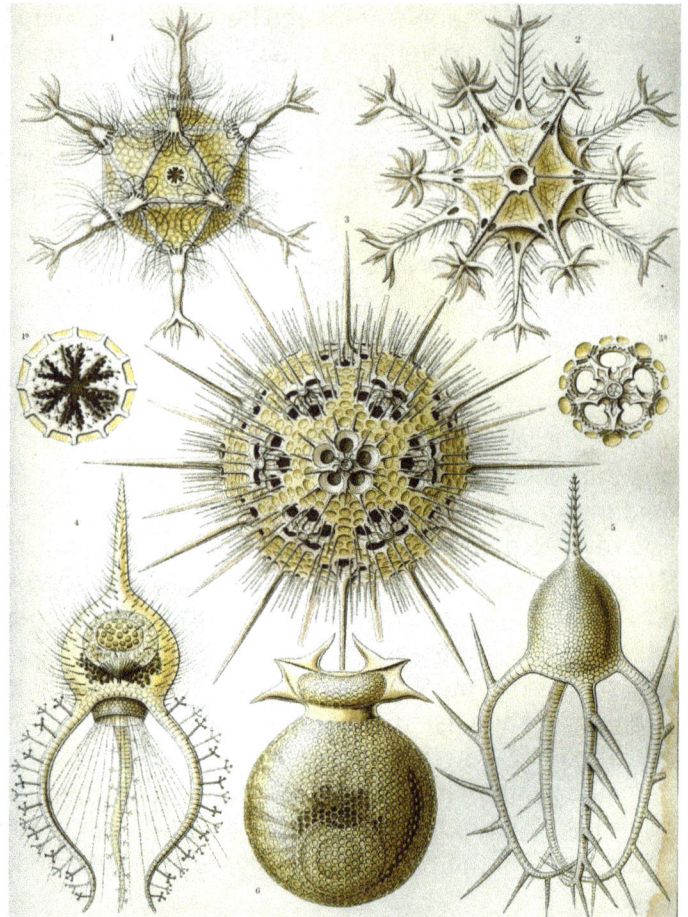

Lastly, we have the **Radiolarians** *(RAY-dee-o--LARE-ee-ans)*. If our little friends hadn't lost their net and had been able catch some of these and bring them back, just imagine how much they could have sold them for! Are these awesome or what? It's hard to believe they are living creatures. Many of these were first discovered on a famous scientific voyage, the Challenger Expedition, in 1873-6. Most were dredged up from the ocean floor, with only their skeletons remaining. Radiolarian skeletons and foraminiferan tests are found by the billions in ocean floor sediments (ofen called "ooze").

Radiolarians are unique among the ameboids in that they share something in common with the diatoms: they can take silicon out of the water. The diatoms use the silicon to build outer shells, with their cytoplasm completely inside. The radiolarians, on the other hand, use the silicon to build glass skeletons as a framework for their ectoplasm to ooze around on. All those fancy spiky things have a thin layer of cytoplasm around them. Hard to imagine, yes. Seems more like science fiction than science. In other respects, radiolarians function pretty much the same as the other ameboids. Same organelles (nucleus, Golgi bodies, mitochondria, etc.) and same method of eating (catching things with the pseudopods then bringing them inside).

Radiolarians like to be at the surface of the water, unlike forams who don't mind being at the bottom. (Although there are a few species of radiolarians that are found at great depths.) To keep themselves afloat, the radiolarians make tiny air bubbles and oil droplets in their ectoplasm. One reason they like to be at the surface is that they often have algae cells living inside of them. The algae cells are called **endosymbionts**, from the words "endo" and "symbiosis." It is not unusual to find dinoflagellates living inside radiolarians. The species of dinoflagellates that become endosymbionts are called **zooxanthellae** *(ZOO-zan-THELL-ee)*. These zooxanthellae are also found living in coral polyps in coral reefs. They help to give corals their intense colors.

Radiolarians show up in great abundance in the fossil record. It is estimated that 90% of all radiolarians are now extinct. As with forams, radiolarians are used by oil prospectors to help them determine where to drill for oil. Certain types of oil-bearing rocks contain certain species of extinct radiolarians.

What is bright yellow, looks like a disgusting clump of something unmentionable, and slowly creeps across the ground? Okay, probably an alien from someone's science fiction story, but besides that, what they have discovered here is a **slime mold**. It got this name before its cellular structure was known. It can make spores like a mold, so it was originally classified as a mold. In recent years we have learned a lot about this form of life, and it is now classified as an ameboid.

The scientific name for this protist is Fuligo septica.

Yet another poster by Ernest Haeckel

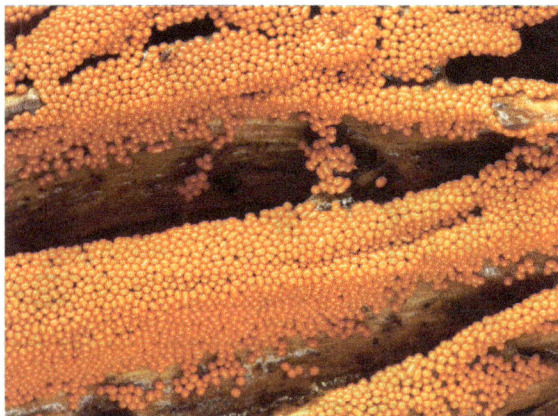

Slime molds come in many colors and shapes.

You can see why our friends thought they had found a very strange clump of poop. Back in the Middle Ages, when even less was known about science, people used names like "witch's butter" or "wolf milk."

Slime molds start out as individual ameboids. When observed as individuals, they have no problem being classified as ameboids. All the things we learned about amebas would apply. It's when they get overfed that they get strange. Give them an abundant food supply (bacteria is a favorite) and they morph into a blob. The individual cells send out signals to each other to swarm and they all join together to make a mega-cell. They actually fuse together so that the mega-cell, called the **slug**, is one giant cell with thousands or millions of nuclei. The slug moves along, gobbling up all food in its path.

Slime molds can move at a rate of about 1 mm per hour (although there are claims that some have been clocked at 1 mm per second). The giant ameboid "slug" moves in the same way an ameba does, by changing the arrangement of the fibers of its cytoskeleton. Once the cytoskeleton changes shape, the cytoplasm just goes along for the ride.

It doesn't take long for the slime mold slug to decide to reproduce. Some parts of it begin to make spore-producing stalks, like mushrooms and molds do. This poster shows some different types of stalks. The tip releases millions of spores that go flying off into the air. If a spore lands in a suitable place, it will grow into an individual ameboid cell. In some species, the individual cells can grow super large, making thousands of copies of their nuclei.

Slime molds can exhibit surprisingly animal-like behavior. Researchers say they have a very basic type of memory that allows them to avoid places that they found unpleasant. If you cut them up, the pieces will find each other and reunite. They may even be able to anticipate environmental changes and prepare accordingly. (As you might guess, preparation would likely start with individuals trading DNA in order to make tough zygotes.)

Next time you take a walk in a forest, see if you can find a slime mold. (Do an Internet search ahead of time and look at some photos.)

ACTIVITY 5.1   ZEUS versus JUPITER, finale

Jupiter left the book in the middle of this chapter, probably right after that bit about the Parthenon. In his opinion, this chapter had way too many Greek stories. Well, as long as Jupiter isn't around, we might as well give you another piece of Greek trivia:

The word "ostrakon" is Greek for "shell." The English word "ostracize" means to ignore someone so that they feel socially abandoned. In ancient Greece, the citizens had the right to vote to banish any politician or leading citizen who got too powerful or dangerous. They would cast their votes by writing on pieces of shell or pottery. (Paper had not been invented yet.) So the shells (ostrakons) were used to vote someone out of town. They were "ostra-cized."

Let's go ahead and finish the word roots, even though Zeus wins by default.

GREEK

amoibe (_____) __ameba__
(am-o-ee-buh)

gaster/gastro (_____) _____

helios (_____) _____

ostrakon (_____) _____

pseudes (_____) __pseudopod___

podos (_____) __pseudopod___

pherein (to bear) _____, and _____

LATIN

rota/roti (_____) _____

foramen (_____) __Foraminiferans___

Greek tends to be more common than Latin in the world of biology. Latin might be used more frequently in other sciences, but Greek seems to be preferred by biologists.

ACTIVITY 5.2   Review questions

1) Things that don't have a definite shape are called **amorphous** ("a" means "without," and "morph" means "shape.") Which of these is amorphous?   a) ameba   b) rotifer   c) gastrotrich   d) radiolarian

2) Which of these carries eggs on her back?   a) rotifer   b) gastrotrich   c) ostracod   d) foraminiferan

3) Which of these was called the "little sun animal" when first discovered?
   a) ameboids   b) rotifers   c) ostracods   d) heliozoans   e) radiolarians

4) Which of these words came from Greek?   a) ameba   b) Stentor   c) heliozoan   d) all of these

5) Which of these is found in abundance in the fossil record?   a) ameba   b) rotifer   c) radiolarian

6) What is a "test"? (Hint: look on page 55)   a) hard shell   b) food grinder   c) egg that survives winter

7) Some ameboids build shells. What do they use?
   a) silicon taken from the water   b) sand and dirt   c) microscopic sea shells

8) Ostracods are almost always: a) male   b) female   c) hermaphrodites   d) none of these

9) Only one of these is a protist. Which one? a) ostracod   b) rotifer   c) gastrotrich   d) slime mold

10) What is an "endosymbiont"?   a) an animal that does photosynthesis   b) an animal that eats algae
   c) an animal that lives inside another animal but does not harm it   d) an animal that makes a hard shell

11) Which of these has a glass shell? a) diatoms   b) foraminiferans   c) heliozoans   d) both a and b

12) Which of these does NOT have pseudopods? a) ameba   b) heliozoan   c) rotifer   d) shelled ameba

13) Which of these is an ostracod most similar to? a) octopus   b) crab   c) worm   d) coral   e) insect

14) A heliozoan is a: a) herbivore, eating algae   b) predator, eating other protists
   c) scavenger, eating dead things   d) photosynthesizer, using sunlight to make its food

15) What does an ameba eat?   a) algae   b) protists   c) bacteria   d) rotifers   e) all of these

ACTIVITY 5.3   Silly nicknames

If protists gave each other nicknames, what would they be? Can you figure out which nickname might go with which protist? The possible answers are:

a) ameba   b) shelled ameba   c) rotifer   d) gastrotrich   e) ostracod

f) heliozoan   g) foraminiferan   h) radiolarian   i) slime mold

1) "Sunny" ___   2) "Crabby" ___   3) "Helmet" ___   4) "Gutsy" ___   5) "Slugger"

6) "Rocky" ___   7) "Wheely" ___   8) "Blobby" ___   9) "Oozy" ___

ACTIVITY 5.4   Decipher these riddles

Here are a few protist riddles. Fill in the answers to these questions, then put the correct letters into the numbered spaces in the answers to the riddles.

1) An organism that has both male and female features is a __ __ __ __ __ __ __ __ __ __ __ __ __
                   17   6   15

2) Females hatching eggs without any male contribution is: __ __ __ __ __ __ __ __ __ __ __ __ __
              1   18

3) This little animal has a foot, stomach, throat (mastax), mouth, muscles and brain. __ __ __ __ __ __ __
                   11   5

4) An organism that lives inside another organism is called a __ __ __ __ __ __ __ __ __ __ __ __
            9   2

5) Slime molds reproduce using these. __ __ __ __ __ __
       4

6) In the Greek story about Hermaphroditus, the nymph used a __ __ __ __ __ to accomplish her goal.
         7   12

7) Amebas move using __ __ __ __ __ __ __ __ __ __.
     3   8

8) Ostracods belong to the class of animals called __ __ __ __ __ __ __ __ __ __.
       16   10

9) Both diatoms and radiolarians can take this element out of the water. S __ __ __ __ __ __
       13   14

****

RIDDLE #1:   What did the foram teacher give his students? __   __ __ __ __
                  1   2 3 4 2

RIDDLE #2:   What holiday can ostracods never celebrate? __ __ __ __ __ __ __ ' __   __ __ __
                    5 1 2 6 3 7 4   8 1 9

RIDDLE #3:   What do amebas use to decorate for Halloween? __ __ __ __ __ __ __ __ __ __ __ __ __
                10 9 2   11 4 12 3 13 3   2 11 14 4

RIDDLE #4:   Why don't diatoms throw stones?

__ __ __ __ __   __ __ __ __ __ __   __ __ __   __ __ __ __   __ __   __ __ __ __ __.
  2 6 3 15 7   6 11 16 4 3 4   1 7 3   17 1 8 3   11 5   18 13 1 4 4

60

EPILOGUE: THE DARK SIDE OF PROTOZOA

Epilogue comes from "epi" meaning "outside," and "logos" meaning "word." So an epilogue, by definition, is "outside (or apart from) the words (the book)." Epilogues are little extra bits that the author thought might be interesting to the reader, but did not want to include as part of the main text. Since epilogues are not really part of the main book, this section is optional. If you don't want to read about yucky protists that make you sick, you can quit reading right now.

Our fingerprint friends ran into some protists armed with weapons, but no one really got hurt. In real life there some protists that can make you very sick, or even kill you. The protozoan that makes the largest number of people sick is the one that causes malaria. In 2013, 198 million people came down with malaria and 584,000 of them died. Most of these cases were in Africa, but malaria is also a problem in India, southern Asia and northern South America. These places are relatively close to the equator. Because malaria is transmitted by a mosquito, it only occurs in areas that are warm year round. Cold weather helps to keep mosquito populations low.

The scientific name for this deadly protozoan **Plasmodium falciparum**. The Plasmodia, as we will call them (since there are several species besides the *falciparum*) used to be classified in a phylum called the Sporozoans. Many books still use this term, and many scientists still use it, too. However, officially, this phylum has been erased and replaced by several other new phyla with terribly difficult names and hard-to-understand definitions. Everything was fine until they started learning too much about these creatures. Worse yet, they probed into their DNA and started doing DNA comparisons. Details that might not sound like a big difference to us are a huge deal to these **taxonomists** (people who classify), and they felt they had to reclassify lots of organisms. In the process, they even got rid of kingdom Protista.

Yes, we've been using that term all through this book, knowing it was "wrong" according to the "highest experts" on classification. Also, there's no animal kingdom anymore. Or fungi. Or plants. In fact, they don't like the word "kingdom" and have abolished it completely. And worse yet, here's a sample of the words they've chosen as replacements: Chromalveolata, Opisthokonta and Archaeplastida. No wonder most textbooks still use the old categories! Classification is a human invention, and in the end it doesn't really matter what names you use; the organisms don't care what we call them.

So back to Protists and Sporozoans...

Plasmodia are still called Sporozoans in many textbooks. On Wikipedia they are listed as kingdom Chromalveolata, phylum Apicomplexa. So we should really call them Apicomplexans. The organisms in this group don't have cilia or flagella or pseudo-pods. They move by doing something called "gliding" but it's not the same kind of gliding that diatoms do. How the Apicomplexans glide is hard to understand, and researchers are still learning about it. Their limited ability to move is mostly directed towards the goal of getting into cells. They are smaller than cells and can live inside of them like houses. They are able to "glide" right in and make themselves at home inside a blood cell or liver cell. Once inside, they are terrible house guests. Not only do they not help with chores, but they start making hundreds of copies of themselves, filling the house to the brim. Finally, the cell bursts and all the new little nasties go off to the next part of their parasit-ic adventure. Sometimes they need to spend time inside two different types of animals in order to complete their life cycle.

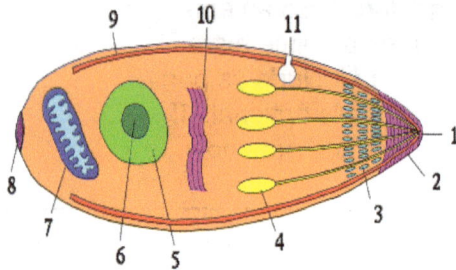

An Apicomplexan: 5 and 6 are the nucleus, 7 is a mitochondria, and 10 is a Golgi body. Parts 1 to 4 help the parasite get into a host cell.

The most famous disease caused by a Plasmodian is *malaria*. Malaria is spread by mos-quitoes. The Plasmodia are in the female mosquito's saliva; when she bites you, some of her infected salvia goes into your skin. It gets into your blood and travels through the bloodstream until it reaches the liver. The Plasmodia cells know to get out of the blood at this point and look for a nice liver cell house. They stay in their liver houses for about two weeks, dividing asexually many times to make thousands of copies of them-selves. The liver cells explode and release all the new Plasmodia into the blood. The Plasmodia then get into your red blood cells. They live there for a while and reproduce not only asexually, but sexually as well, making male and female reproductive cells called *gametes*. When the red blood cells explode and release all the Plasmodia and their gam-

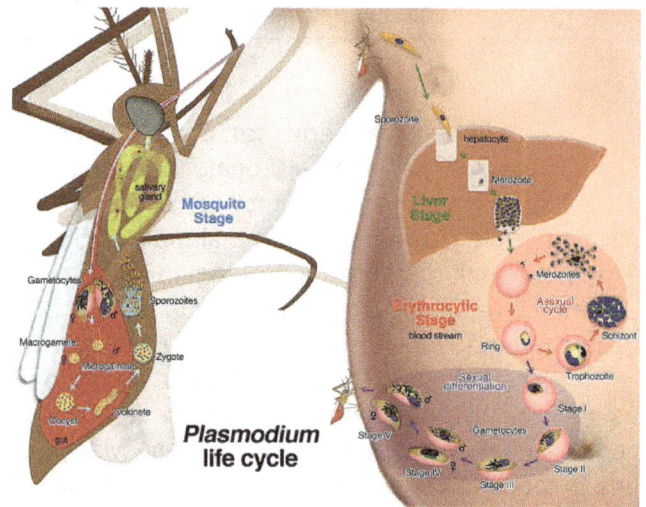

Illustration credit: Wikipedia "Plasmodium malariae"
Graphics by La Roche Lab, UC Riverside (http://ucrtoday.ucr.edu/19520)

etes, you experience the high fevers and chills that are the main symptoms of malaria. At this point, the gametes have to wait, and hope the sick person attracts another mosquito. If a female mosquito comes along and bites the sick person, the mosquito will draw up blood that contains the gametes.

Once inside the mosquito, the gametes procede with the next step. They fuse to form a zygote-type-thing. This cell makes a home in the mosquito's gut and starts producing hundreds of Plasmodia that will then travel to the mosquito's salivary glands, ready to be injected into a new human victim. Round and round the cycle goes, year after year, century after century.

Parts of North America used to have malaria, but intensive use of insecticides in the 1900s managed to put an end to malaria in that part of the world. Why can't we just do the same thing in other parts of the world? Those insecticidal chemicals have been banned, never to be used again. So now it's up to scientists to come up with another way to fight malaria. Perhaps you will find things you can do in your lifetime to help the fight against malaria.

There's one other member of the Apicomplexa group that you need to know about. It's kind of a "cousin" to malaria and is called *Babesia* (*Bab-EE-zee-ah*). It was named after a

Romanian scientist named Victor Babes. (Would you like to have a disease named after you?) *Babesia* gets into your red blood cells, just like malaria does. In North America, it most often occurs as a **coinfection** along with Lyme Disease (which is caused by bacteria). A coinfection is something you catch along with the main thing you catch. It's a bad bonus.

Both the Lyme bacteria and the *Babesia* protozoan are transmitted by ticks. One tick can carry as many as five or six different infections! Most doctors are not yet aware of how common (and how dangerous) *Babesia* is. The number of cases is growing at an alarming rate, and the symptoms don't always look like malaria symptoms. Anyone who suspects they have Lyme Disease should also be tested for *Babesia* and other coinfections.

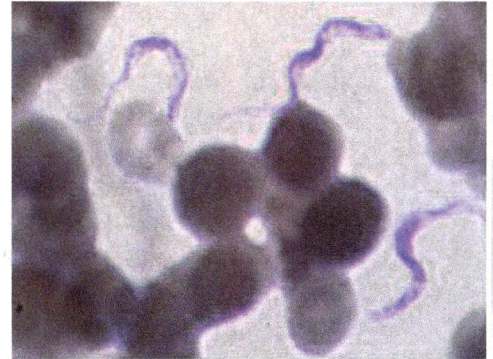

Trypanosomas swimming around blood cells

The next villain on our list is a member of the euglenoid group. The euglena we met in chapter 3 was a very nice one. It was a photosynthesizing protist that helpfully turned sunlight and carbon dioxide into sugars that other organisms, including humans, could use for food. The euglenoid called **Trypanosoma** doesn't do photosynthesis. It survives by living inside other organisms as a parasite. Living inside another organism doesn't automatically make you bad; endosymbionts such as the dinoflagellates (zooxanthellae) that live in coral are beneficial and help to keep the corals healthy. The word "symbio" indicates a beneficial relationship between the organisms. The word **parasite** indicates that the host organism is being harmed in some way. Parasites don't always kill their hosts, and, in fact, it is to their advantage to keep their host alive so they can continue to have a place to live. However, parasites can do a lot of damage and sometimes end up killing the host.

The *Trypanosoma* is a harmful parasite that causes a disease called "sleeping sickness" which kills thousands of people in Africa every year, especially in the Congo region. (A similar disease called Chagas disease afflicts people in Central and South America.) The name of the disease comes from the fact that after you have the initial fever, chills, aches and swollen lymph glands, you begin to have neurological symptoms that affect the brain and spine. One of the brain areas affected is the area that regulates the sleep cycle, and you begin having very strange sleep patterns, often being too sleepy during the day.

The *Trypanosoma* parasite is similar to malaria in that it is transmitted by an insect. African sleeping sickness is carried by the Tsetse *(tzee-tzee)* fly. The *Trypanosoma* lives in the fly's digestive system and makes its way up to the saliva glands, just like the malaria Plasmodium does. When the fly bites someone, the parasite is injected along with the fly's saliva. If the disease is diagnosed early, it can be treated with medicines, but many people don't have adequate medical care available.

The name "*Trypanosoma*" has an interesting story attached to it. The word "trypano" is Greek and means "to drill a hole." In ancient times, doctors did something called **trepanning**. They drilled a hole in the skull as a treatment for various head ailments including headaches and seizures. They mistakenly believed that the cause of these problems was too much pressure on the brain. Opening the skull to reveal the brain was a terrible idea, but surprisingly, many people survived this operation. Trepanning continued to be popular in the Middle Ages and Renaissance. Its popularity tapered off after that, but there are still recorded trepanning surgeries as late as the 1800s.

The *Trypanosoma* doesn't drill into people's heads. It drills into cells. Sometimes it doesn't even have to drill— it can simply touch molecules on the outside of the cell that act like doorbells. The Plasmodium knows how to "ring the doorbell" so that the cell will let it come in. Once inside, the *Trypanosoma* begins to reproduce itself, eventually bursting the cell.

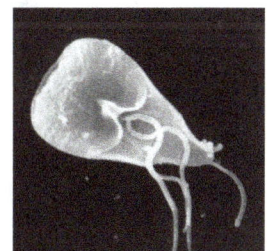

A parasite called **Giardia** *(gee-AR-de-ah)* is one of the main reasons you need to use a water filter if you are out hiking and want to drink from a stream or pond. *Giardia* is a flagellated protist but is not a euglenoid; it's in its own phylum group. (The name of which is not well known, and not worth bothering you with.) It has eight flagella; some of them are visible in this photograph. The place it likes to call home is the small intestine of many mammals, including humans. When not in an intestine, it can be found in ponds, streams, puddles, drainage ditches, and sewers.

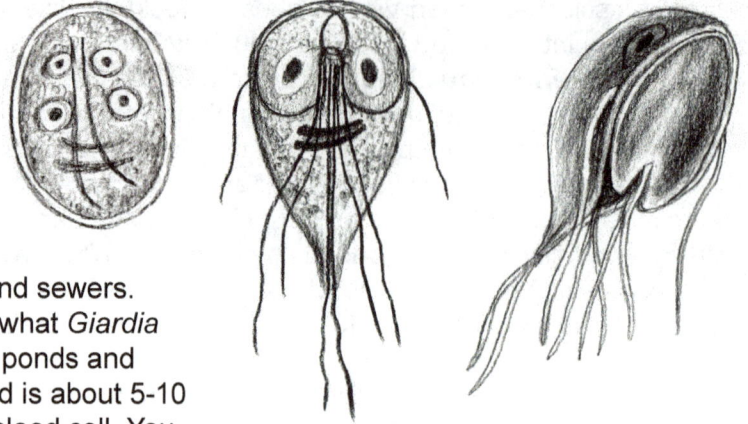

The first picture, on the left, is what *Giardia* looks like when it is floating around in ponds and streams. This form is called a **cyst** and is about 5-10 microns long, about the size of a red blood cell. You can see a protective layer all around it. The cyst often has four nuclei visible. It can wait for months or years until it is swallowed by a mammal. *Giardia* has been found in dogs, cats, sheep, horses, cows, deer and beavers. The middle picture shows what *Giardia* looks like while it is living in a small intestine. It is about 20 microns long. Not many protists are that small. The things that look like eyes are two nuclei. It is unusual for a cell of any kind to have more than one nucleus. Diagnosing *Giardia* using a microscope is very easy because they look like smiley faces! No one is exactly sure what those "mouth" lines are. The last picture shows the suction cup on the bottom side of the cell. The suction cup helps the *Giardia* to stick to the cells that line the intestines and not get washed out by all the food and water that passes through.

Giardia was first discovered by Antoni von Leeuwenhoek in 1681. He was having diarrhea and wondered if some microorganism might be causing it. He put a drop of diarrhea on his microscope and saw what we now call *Giardia*. Leeuwenhoek did not have a name for this organism and it remained nameless until 1895 when Alfred Giard began researching it. Since then it has been found to be the most common intestinal parasite in the world.

Giardia attaches to the wall of the small intestine. It is never found anywhere else in the body, not even in the large intestine. It can cause nausea, vomiting, and fever, and always causes chronic (long-lasting) diarrhea. Viruses that attack the intestines are usually gone in a few days or maybe a week. If you have diarrhea that lasts several weeks or longer, a doctor can take a "stool sample" (a tiny bit of poop) and have a lab technician put it on a microscope slide and look for *Giardia*. If *Giardia* is found, the doctor can prescribe medicines that will kill the parasite. Before modern times, people often treated parasites of the intestines with herbs such as artemesia ("wormwood") or goldenseal.

To avoid catching *Giardia*, don't ever drink unsafe water. *Giardia* can survive chlorine (bleach) so water treatment tablets (which are basically chlorine tablets) won't kill *Giardia*. You must either boil the water for 20 minutes, or use a filter that catches particles as small as 5 microns. Filters should come with a label that tells the smallest particle size they can catch.

Our last nasty beast is called **Entameba histolytica** *(HIST-o-LIT-i-cah)*. "Histo" means "tissue," and "lytic" means "destroying." This is an ameba that likes to destroy your intestines and then, if possible, bore its way through the intestinal wall and get into your other organs. It can end up in the liver where it creates an pocket of disease called an abscess. When your own immune system cells come to fight off the ameba, they also get destroyed and release their powerful chemicals into the area, causing even further damage. This sounds terrible, and it is, but amazingly, *Entameba* infections can be so mild that the host human doesn't know that anything is wrong. An estimated 50 million people worldwide are infected with *Entameba histolytica*.

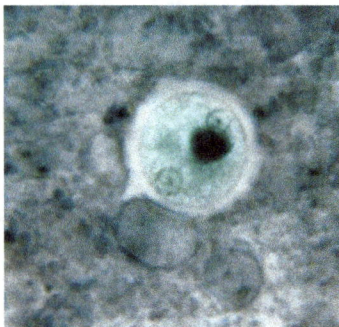

You can prevent an *E. histolytica* infection by making sure your drinking water and food are clean. If you suspect the water might have *Entameba* in

ACTIVITY E.1   Parasite puzzle

tick
genus *Ixodes*

ACROSS
3) Giardia has 8 of these.
4) A tough egg-like form that can survive harsh conditions.
5) The euglenoid that causes African sleeping sickness.
7) This word is very hard to spell and means watery feces.
10) An organism that lives on or inside another organism and causes damage to its host.
11) When a mosquito or a fly bites you, some of this liquid gets transfered into your skin.
13) This insect is the "vector" for malaria.
16) The phylum Sporozoa is now called _____.
19) This fly is the "vector" for African sleeping sickness.
22) An organism that lives inside another organism but provides a benefit to its host.
23) This is what you have to do to pond or stream water for 20 minutes to make sure it does not contain any parasites.

DOWN
1) The malaria parasite's correct name is Plasmodium _____.
2) Most common infection carried by ticks: ____ Disease.
4) Long-lasting (pg. 64)
6) The ancient practice of drilling into a skull.
8) Giardia likes to live in the small _____.
9) A malaria-like parasite that is found in N. America.
12) An infection that you get along with the main one.
14) "Mal" means "bad." This disease was once thought to be caused by "bad air."
15) "Histo" means ____. (pg. 64)
17) A tiny bit of poop is called a _____ sample. (pg. 64)
18) Male and female cells (egg, sperm) are called ____.
20) Many people in the world have used (and still use) plants called _____ to get rid of parasites. (Examples: wormwood and goldenseal)
21) This relative of the spider can carry more than one type of infection, including bacteria and protozoans.

ACTIVITY E.2   Visual quiz

How well do you know your protozoans? Try to match the pictures with the names. These drawings are "not too scale," meaning that large things look the same size as small things.

1) Ameba _____
2) Paramecium _____
3) Plasmodium _____
4) Epistylis _____
5) Volvox _____
6) Ostracod _____
7) Rotifer (Philodina) _____
8) Gastrotrich _____
9) Didinium _____

10) Foraminiferan _____
11) Desmid _____
12) Spirogyra _____
13) Shelled ameba _____
14) Chlamydomonas _____
15) Diatom _____
16) Loxodes _____
17) Radiolarian _____
18) Coleps _____

19) Cyanobacteria _____
20) Dinoflagellate _____
21) Giardia _____
22) Euplotes _____
23) Stentor _____
24) Dileptus _____
25) Euglena _____
26) Vorticella _____

INDEX

The index can help you find something if you can't remember where it was in the book.

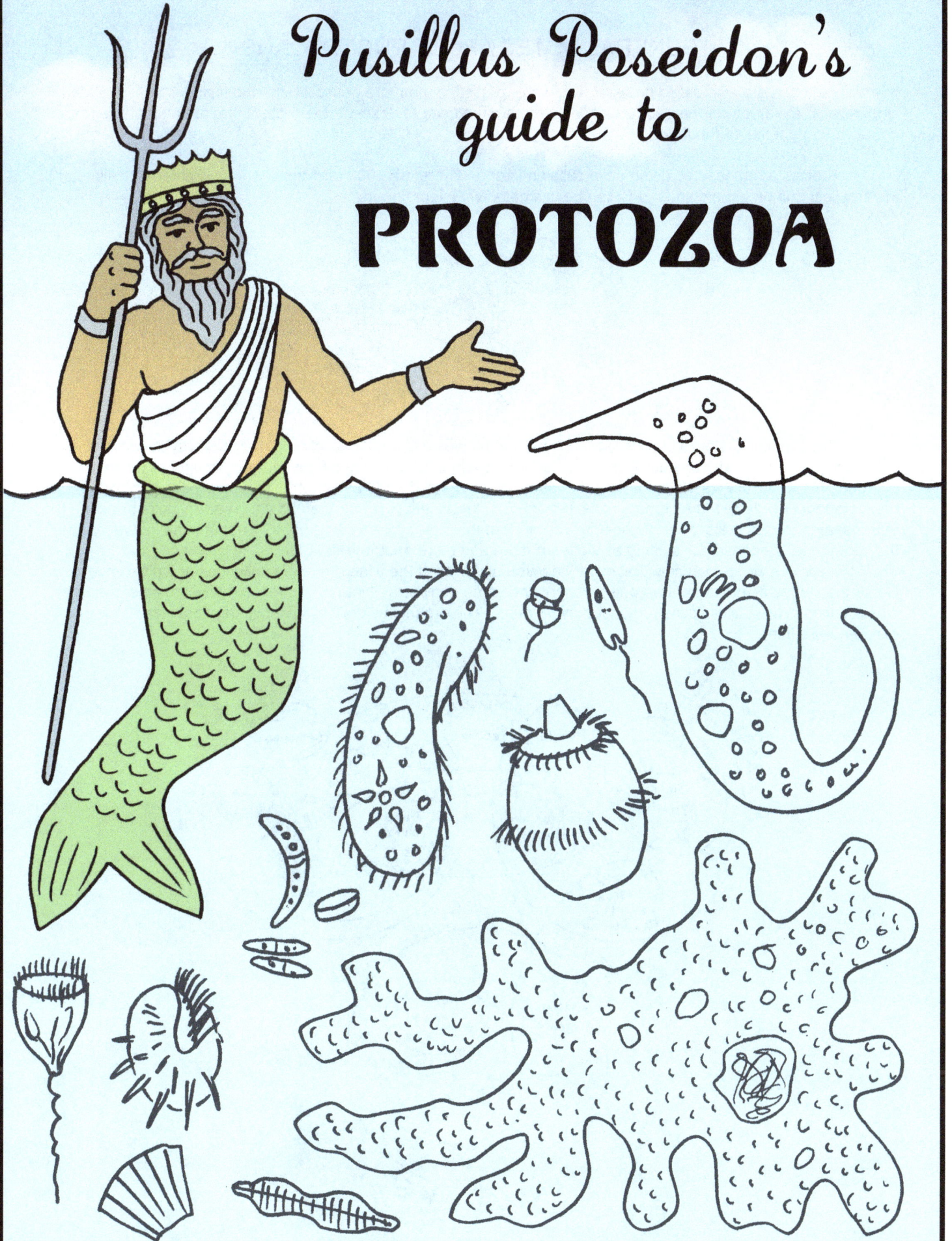

Pusillus Poseidon's guide to PROTOZOA

GENERAL NOTES ABOUT PROTOZOANS

Protozoa are also called protists. The word "protist" is the more general term and includes all types of single-celled eukaryotes, whereas "protozoa" is more often used to describe the protists that are animal-like (as opposed to plant-like or fungi-like).

Protists are measured using units called microns. There are 1000 microns in one millimeter. A millimeter is the smallest unit on a metric ruler and can be estimated with your fingers:

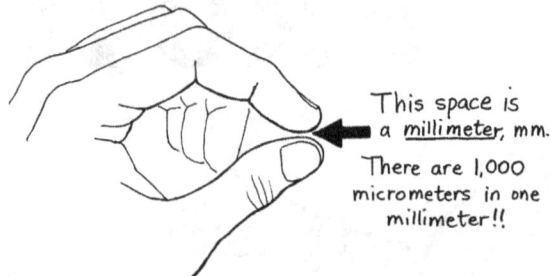

This space is a _millimeter_, mm.

There are 1,000 micrometers in one millimeter!!

The traditional way of classifying protists is by the way they look (morphology), by the way they move (motility), and how and what they eat. This gives us terms such as ciliates, flagellates, ameboids, and all those colors of algae. Recently, the classification system has been overhauled and has become immensely complicated. (Information about DNA is now the primary consideration for classification, rather than how a creature looks or acts.) If you research these creatures on Wikipedia, you will see this new system being used. Bear in mind, however, that the categories are constantly shifting as we learn more and more about protist DNA.

Here is a visual overview that might help you understand the wide range of similarities and differences. Some organisms fit into more than one category and some don't fit well into any category. Always remember that classification is an artificial construct made by humans. The organisms don't know anything about it and they don't care what we think!

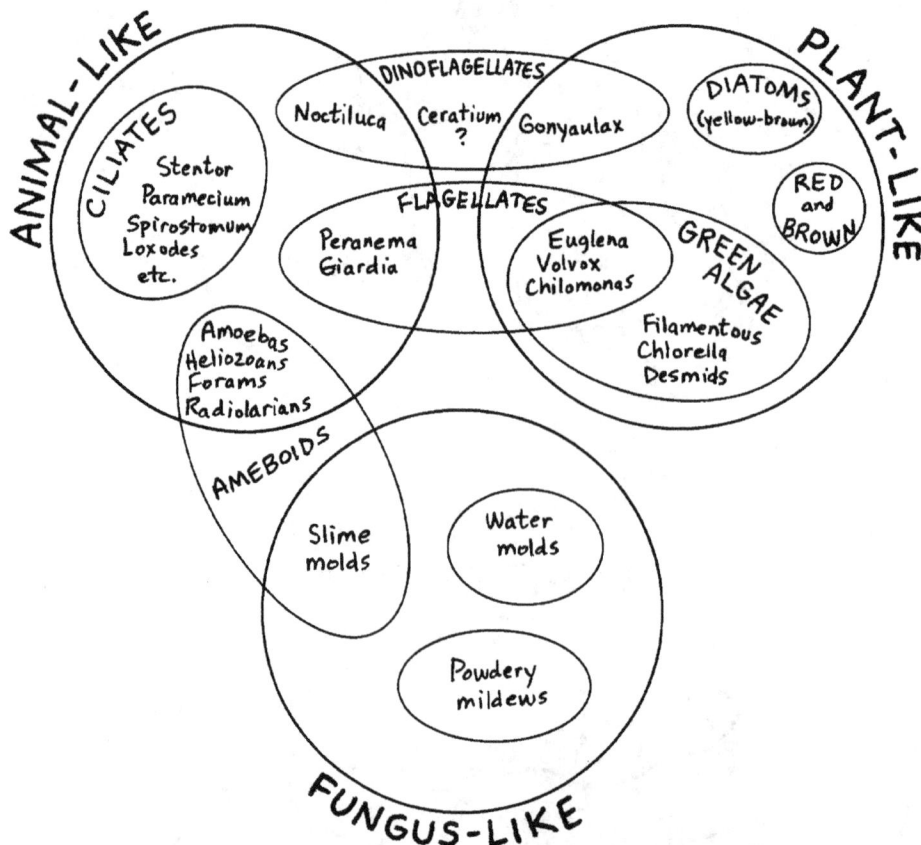

CILIATES

	Blepharisma 150-200 μm	Eats anything smaller than itself, even smaller *Blepharismas*. (Cannibalism causes them to double their normal size. The reason for this is unknown.)	*Blepharisma* looks slightly pink because it makes a red pigment that senses light (similar to the pigment in *Euglena*'s red eye spot). *Blepharisma* hides from light. The cilia are much longer on one side, as you can tell from the picture.
	Bursaria 500-800 μm	Eats smaller ciliates, and particularly enjoys eating the *Paramecium* species *Paramecium bursaria*.	Has a huge gullet opening, large enough to swallow *Paramecia*. When environmental conditions are bad, *Bursaria* can form a tough cyst and "hibernate" in this form until conditions are good again.
	Coleps 50-80 μm	Will try to eat just about anything. Is equipped with trichocyst darts that it can use to paralyze prey. Will take "bites" out of large prey (even ostracods).	Has armor plates made of calcium carbonate, the same material that clam shells are made of. Has thick barbs at one end.
	Colpidium 50-70 μm	Eats bacteria.	Can reproduce by binary fission every 4-6 hours. Often used for experiments in labs. *Colpidium* can survive in polluted water so the presence of many *Colpidium* might indicate poor water quality.
	Didinium 100-200 μm	Eats other ciliates, especially *Paramecia*. (*Didinia* can digest 2 *Paramecia* per hour.)	*Didinia* have trichocysts to help it capture prey. Like *Blepharisma*, a *Didinium* can "hibernate" by making itself into a cyst (a tough little lump that will not dry out).
	Dileptus 200-400 μm	Eats other ciliates and also rotifers.	*Dileptus* has a proboscis lined with toxins. It can stun its prey by hitting it with the proboscis. The mouth of a *Diletpus* is located at the base of the proboscis.
	Epistylis 50-150 μm	Eats small particles and bacteria that it sweeps in with its top ring of cilia.	Lives in colonies. It might look like *Vorticella*, but *Epistylis* can't spring up and down because it does not have myonemes, the fibers that act like muscles. *Epistlylis* doesn't move much. Colonies are often attached to plants, shells or insects.
	Euplotes 80-100 μm	Eats anything smaller than itself.	Euplotes has long cirri on its "bottom" end, which it can use not only for swimming but for walking along surfaces.

	Halteria 25-50 μm	Eats bacteria, algae and tiny protists.	Halteria looks like it is jumping around as it moves. This jumping motion is caused by the groups of long cilia on the middle of its body.
	Loxodes 700 μm	Eats bacteria, algae and small protists.	*Loxodes* is known for its ability to sense which way is "up." This ability comes from the "Müller body" structures along its dorsal (back) side. (The Müller bodies contain a crystal of barium sulfate.) *Loxodes* has a rostrum (beak-like structure).
	Paramecium bursaria 100-200 μm	Eats mostly bacteria and algae, but also gets some nutrition from the algae that live in its cytoplasm as endo-symbionts.	*Paramecium bursaria* is the only species of paramecium that has a symbiotic relationship with algae. It takes in *Zoöchlorella* algae cells that then live in its cytoplasm, doing photosynthesis. Both organisms benefit from the sugars the algae makes.
	Paramecium caudatum 200-250 μm	Eats bacteria, algae and small protists.	*Paramecia* have weapons called trichocysts that function like tiny harpoons. The darts contain a toxin that can stun or partially paralyze attackers or prey.
	Podophyra 10-30 μm	Sucks cytoplasm out of other protists, especially ciliates.	The "hairs" sticking out are actually sucking tentacles that can suck cytoplasm out the ciliates it preys upon. Often, *Podophyra* will not take all the cytoplasm and will release the ciliate alive. *Podophrya* is sessile (does not swim).
	Spirostomum 800-1000 μm	Eats bacteria, single-cell algae and tiny bits of debris. The mouth is very small.	When startled, *Spirostomum* contracts quickly to less than half its length. The contraction is caused by myonemes, fibers that act like muscles. This contraction is the fastest cell movement in the world, as far as we know. (*Vorticella* does it, too.)
	Stentor 500-2000 μm	Eats anything it can sweep into its gullet.	The ring of cilia on the top moves to create a current that will bring food particles in toward the gullet opening (mouth). *Stentor* can attach itself to a surface, or it can swim around freely. They often look green, due to symbiotic algae living inside them.
	Stylonychia 150 μm	Eats anything smaller than itself.	One of the most common ciliates, as common as *Paramecium*. The long, thick, cirri help it to move along surfaces, similar to the way *Euplotes* moves.

CILIATES, continued

	Tetrahymena 40-80 μm	Eats bacteria and tiny particles.	This is one of the most studied ciliates. It is commonly used in labs to do genetics studies and physiology experiments because it is easy to grow and won't die easily.
	Vorticella 50-150 μm	Eats anything small enough to get swept into its gullet by the ring of cilia.	Feeds in a manner similar to *Stentor*. When startled, it contracts its stalk. The stalk contains myonemes that act like muscles fibers (similar to those you find in *Spirostomum*). *Vorticella* like to live in groups, but they <u>don't</u> actually connect like *Epistylis*.

AMEBOIDS

	Amoeba proteus 300-600 μm	Eats anything it can catch.	Feeds by using pseudopods, which are long extensions that it "oozes" out. The pseudopods close around the food, creating a vacuole that it brings inside. The ameba has no definite shape and changes all the time. Its name means "changeable changer."
	Entamoeba histolytica (hist-o-LIT-i-cah) I5-60 μm	Lives inside the intestines of humans and other mammals. Eats bacteria and small food particles.	One of the most common parasites in the world. Infections can be so mild that people are even unaware they have it. Symptoms include diarrhea, weight loss and fatigue. The parasite is most often contracted by drinking dirty water.
	Heliozoan 200-300 μm	Eats anything it can catch with its long axopods. (Prey is usually small.)	When first discovered, they were called "sun animacules" because they looked so much like a little sun with rays shining out. Their axopods are used not only for capturing food but also for sensing the environment and for attachment to surfaces.
	Foraminiferan I00-1000 μm	Eats anything it can capture with its long axopods. (Prey is usually small.)	Foraminiferans (or "forams") make shells out of calcium carbonate, the way clams do. There are thousands of different kinds of forams, some large enough to see without a microscope. Some species live in the deepest part of the ocean (the Mariana Trench).
	Radiolarian I50-300 μm	Eats anything it can capture with its pseuodopods and and/or axopods (depending on species). Prey is usually small. Some speciers have dinoflagellates living inside.	Radiolarians can take the element silicon out of the water and use it to make complex "skeletons" made of silicon dioxide (glass). There is a thin layer of cytoplasm that flows around the outside of the glass skeleton. Radiolarians come in many different shapes.

AMEBOIDS, continued

	Arcella (Ar-SELL-ah) 30-100 μm	Eats anything it can catch with its pseudopods, usually diatoms, algae, and tiny protists.	*Arcella* is a shelled ameba. The shell is called a "test" and is made out of a protein called chitin.
	Difflugia (Dif-FLU-gee-ah) 100-150 μm	Eats anything it can catch with its pseudopods	This is a shelled ameba. Makes a shell ("test") out of grains of sand and tiny bits of debris and dirt.
	Fuligo (yellow slime mold) Colonies can be the size of a plate.	In its cellular form, the individual amebas eat mostly bacteria. In its "slug" from, it gobbles any small organic particles in its path.	*Fuligo septica* is a bright yellow slime mold. In its "slug" form (millions of individuals all joined together) it can slowly creep along.
	Physarum (Fi-SARE-um) Colonies can be the size of a plate.	Eats bacteria, fungal spores, and other microorganisms.	Physarum is a large group of slime molds, with many different species. They are found in cool, moist places like damp forests that have many decaying logs.

SINGLE-CELL FLAGELLATED GREEN PROTISTS

	Chilomonas 20-40 μm	Uses sunlight for photosynthesis.	*Chilomonas* is a very common algae, found in fresh water and ocean water. *Chilomonas* is a member of the "cryptomonads" group, who have special structures near the flagella that allow them to jet away from irritants very quickly.
	Chlamydomonas 10-30 μm	Uses sunlight for photosynthesis.	*Chlamydomonas* is found in many places including fresh water, ocean water and in damp soils. Chloroplast is bowl-shaped.
	Phacus 40 μm	Uses sunlight for photosynthesis.	Is flat and leaf-shaped. Has small chloroplasts (unlike Chilomonas and Chlamydomonas which have a very large one). Found in fresh water ponds. Has a red eye spot.

SINGLE-CELL FLAGELLATED PROTISTS (not necessarily green)

	Euglena gracilis 100-200 µm	Can make its own food using photosynthesis. However, if conditions are poor for photosynthesis, it can also catch and eat bacteria and small protists.	Has a red eye spot that helps it go toward sunlight. The red spot itself does not detect light, but filters light onto a light sensitive spot. The end of the body with the flagella can also engulf prey and bring it inside. This species is often used in laboratories.
	Euglena acus 50-175 µm	Can make its own food using photosynthesis. However, if conditions are poor for photosynthesis, it can also catch and eat bacteria and small protists.	*Euglena* do not have a cell wall like algae do. They have a flexible pellicle. *Euglena* are very nutritious and can be used as a powdered food supplement.
	Peranema 20-70 µm	Eats bacteria, algae and small protists. (It captures and eats its prey in a manner similar to *Euglena*.)	Though it is similar to *Euglena* in some ways, *Peranema* does not have chloroplasts and can't do photosynthesis. It is easy to identify because it keeps its flagella very straight and only wiggles the very end.

SINGLE-CELL GREEN PROTISTS

	Chlorella 5-10 µm	Uses sunlight for photosynthesis.	Is one of the smallest but most common types of algae. Can be found anywhere there is moisture, even on damp trees and rocks. It is very nutritious and is used as a powdered food supplement.

DESMIDS (single-cell green protists that look like diatoms but are actually green algae)

	Closterium 100-400 µm	Uses sunlight for photosynthesis.	Has a vacuole at each end. (The lines of dots are not vacuoles but are involved in photosynthesis.) Movement is restricted to slowly pivoting, by pushing fluid out of pores (similar to diatom movement). It usually green or yellowish-green.
	Cosmarium 30-70 µm	Uses sunlight for photosynthesis..	Looks like two cells, but is actually just one cell with two halves. The narrow middle part is often called the isthmus. Other species of desmids have more intricate shapes.

FILAMENTOUS GREEN PROTISTS

	Oedogonium (Oh-do-GO-nee-um) 10-20 µm wide	Uses sunlight for photosynthesis.	Specialized Oedogonium cells swell up and produced an egg. Other cells make sperm. The eggs and sperm fuse to form zygotes.
	Spirogyra 35-45 µm wide	Uses sunlight for photosynthesis.	Known for its spiral-shaped chloroplasts. The nucleus is suspended in the middle of the cell. In ponds, mats of green scum are often made of Spirogyra.
	Vaucheria (Vow-chair-ee-ah) 40 µm wide	Uses sunlight for photosynthesis.	Life cycle is similar to Oedogonium. Very large bumps are clearly visible when viewed under the microscope.
	Zygnema 20-70 µm wide	Uses sunlight for photosynthesis.	Chloroplasts are star-shaped. Life cycle is similar to Spirogyra.

COLONIAL GREEN PROTISTS that live in spherical colonies

	Gonium (Colonies can be up to100 µm.)	Uses sunlight for photosynthesis.	Colonies stay small, only 4-16 cells. The colony is flat (unlike *Volvox*). A thick gel keeps the cells together. Colonies can reproduce by splitting in half, or some cells can specialize to make eggs and sperm that will form a zygote to start a new colony.
	Pandorina (Colonies can be 50-250 µm.)	Uses sunlight for photosynthesis.	Colonies stay fairly small, usually no more than about 32 cells. Each cell has two flagella and an eye spot for detecting light. Colonies can split in half to form a new colony, but can also use sexual reproduction with eggs and sperm.
	Volvox (Colonies can be up to 1000 µm.)	Uses sunlight for photosynthesis.	*Volvox* colonies can grow very large, with hundreds of cells. The colony forms a sphere. Each cell has two flagella and the cells cooperate to move the colony around. Some cells specialize to make eggs or sperm.

DIATOMS (yellow-brown protists)   Note: They can do photosynthesis even though they are not green.

	Asterionella Each cell is 40-80 µm long	Uses sunlight for photosynthesis.	Their name means "little star." Each arm of the star is one cell, and they are all joined at the center. Colonies consist of 8-20 cells.
	Fragilaria Each cell is 40-60 µm long (width of colony)	Uses sunlight for photosynthesis.	Cells are joined in the middle to form long colonies that look like a row of books on a shelf. There are many different kinds of *Fragilaria*, living in both fresh water and ocean water.
	Gomphonema 20-80 µm	Uses sunlight for photosynthesis.	Has H-shaped chloroplast. Can tolerate water rich is nitrates and phosphates (considered to be pollution), so the presence of many *Gomphonema* might be an indication that the water is becoming polluted.
	Meridon Each cell is 40-50 µm long (width of colony)	Uses sunlight for photosynthesis.	To the naked eye, *Meridon* would look like a brown scum on the bottom of a ditch or puddle. *Meridon* is often found in water that contains many cyanobacteria.
	Navicula 100-200 µm	Uses sunlight for photosynthesis.	The name means "small boat." You can see that the shape resembles the hull of a boat. There are over 1,000 different species of *Navicula* diatoms.
	Stephanodiscus 60-70 µm dia.	Uses sunlight for photosynthesis.	Very common. Found floating freely, not attached to anything. The view here is from the top. The actual shape is a very short cylinder, like a petri dish used in a lab. Sometimes they look like they have little threads hanging of the edges.
	Synedra 100-120 µm	Uses sunlight for photosynthesis.	These diatoms are similar to *Fragilaria* except that they do not join together to form colonies.
	Tabellaria Each cell is 40-50 µm long.	Uses sunlight for photosynthesis.	The name means "little tablets." A small group of cells makes a tablet, then these tablets stay joined at one corner. Often sticks to plants or rocks, and does not mind slightly acidic water.

CYANOBACTERIA (do photosynthesis, fix nitrogen, and used to be called blue-green aglae)

	Anabaena (An-ah-BANE-ah) Filaments are 8-10 μm wide.	Uses sunlight for photosynthesis.	Not only can this bacteria do photosynthesis, it can also "fix" nitrogen, which makes it important for nearby plant life. (Those circles are where the nitrogen is processed.) Unfortunately, it also makes a toxic chemical that is harmful to people and animals.
	Nostoc Filaments are 5-10 μm wide.	Uses sunlight for photosynthesis.	This species looks very much like *Anabaena*, and is a member of the same family. However, *Nostoc* can grow colonies so large they can be seen without magnification. The colonies look like leaves and hollow berries lying on the ground.
	Merismopedia Cells are 1-3 μm.	Uses sunlight for photosynthesis.	These bacteria are found in both fresh water and salty water. They produce mild toxins that can irritate skin. The cells reproduce in such a way that they form flat sheets held together by a gelatinous substance.
	Gloeocapsa (Glo-ee-o-cap-sa) Cells are 10 μm	Uses sunlight for photosynthesis.	The name means "glue box." This bacteria is responsible for all those black streak stains you see on shingled roofs. The bacteria feed on the limestone that is used to make fiberglass shingles. The black color is the bacteria's protection against UV rays.

DINOFLAGELATES

	Ceratium (Sare-ah-SHE-um) 100-500 μm	Can do photosynthesis or capture small particles of food (similar to Euglena).	Found in both fresh water and ocean water. They have two flagella but move slowly because of their long "arms." They can reproduce quickly to form "blooms" but are not nearly as toxic as other species of dinoflagellates.
	Gonyaulax (Go-nee-ALL-ax) 30-50 μm	Uses sunlight for photosynthesis.	Mostly found in ocean, with only a few fresh water species. These are the nasties that cause the toxic red tides. Their toxins can accumulate in ocean animals, causing humans to get sick if they eat them. The name means "knee with a furrow/rut."
	Noctiluca 200-2000 μm	Eats anything it can catch, including bacteria, diatoms, small protists, fish eggs and other dinoflagellates.	Many dinoflagellates glow in the dark to some degree, but this is the one that is famous for it. *Noctiluca* means "night light." The thing that looks like a flagella is actually a tentacle used for feeding. It does have a flagella but it is hard to see.

ANSWER KEY

ANSWER KEY:

Chapter 1
Activity 1.2:
Pellicle is like skin.
Cilia is like hair.
The oral groove is like a mouth.
The gullet is like a throat.
The anterior is the head.
The contractile vacuole is like a kidney, getting rid of excess water.
A food vacuole is like a stomach, holding food that is being digested.
Trichocysts have no equivalent in the human body.
The trichocyst launching mechanism is similar to contraction of muscles because it uses calcium.
The lungs have no equivalent.

Activity 1.3:
1) multicellular   2) oblong   3) slippers   4) pellicle
5) plasma membrane   6) anterior   7) diffuse   8) cilia
9) contractile vacuole   10) micronucleus   11) binary fission   12) 200
13) bacteria, algae, yeast cells, rotting debris, small ciliates   14) mouth
15) hair   BONUS: mitochondria   SECOND BONUS: Golgi body

Chapter 2
Activity 2.2:
The peduncle is like a foot.
The cilia can function as fingers because they gather food. (Previously we matched them with hair.)
The dorsal side is like our back.
The buccal cavity is like a mouth.
A myoneme functions like a muscle fiber.
A contractile vacuole is like a kidney because it filters and expels water.
A food vacuole is like a stomach because it stores and digests food.
The Müller bodies function like the balance sensor in the inner ear.
The rostrum looks like a nose.

Activity 2.3: ("Who Am I?")
1) Stentor   2) Spirostomum   3) Didinium   4) Euplotes   5) Halteria
6) Vorticella   7) Epistylis   8) Loxodes   9) Dileptus   10) Loxodes
11) Paramecium   12) Stentor   13) Dileptus   14) Tetrahymena   15) Spirostomum

Activity 2.4:
GREEK:
boskein (to feed): proboscis
di: dileptus, didinium
dinos: didinium
epi: Epistylis
eu (good): Euplotes
halteres: Halteria
hymen: Tetrahymena
kolpos: Colpidium
pro (for): protozoa, proboscis
speira: Spirostomum
stoma (mouth): Spirostomum, cytostome
stulos: Epistylis
tetra (four): Tetrahymena
toxon: toxicysts

LATIN:
bucca: buccal
carn (flesh): carnivore
dorsum (back): dorsal
pedis (foot): peduncle
rostrum (beak): rostrum
sessilis: sessile
spira: Spirostomum
vorare (to eat): carnivore
vortex: Vorticella

Chapter 3

Activity 3.1:

Note: We've only listed the words from this chapter. Adding other words (such as gyroscope) is fine!

GREEK:
auto: (self) autotroph
trophe: (nourishment) autotroph, heterotroph
hetero: (different) heterotroph
chloro: (green) chlorophyll, chloroplast
eu: (good/well/true) Euglena, eukaryotic
glene: (eye) Euglena
gyro: (turn) Spirogyra
pro: (before) protist, prokaryotic
zygo: (yoke) zygote, zygospore

LATIN:
con: (with) conjugation
filum: (thread) filamentous
jugo (join) conjugation
pro: (before) protist, prokaryotic

Activity 3.2:

1) A zygote is the joining of egg and sperm. It's like they are yoked (held tightly) together as a team. Only this team is more permanent than a ox team! Once joined, the egg and sperm never separate but actually lose their individual identity and form a new cell that will grow into a unique organism.

2) Euglena

3) Chlorella does not need to eat like a Paramecium does because Chlorella does photosynthesis with its chloroplasts.

4) Eutrophic means "good nutrition" or "plenty of nutrients." It sounds like a good thing, right? Plenty of food for things that live in the pond. However, the word eutrophic is usually used to describe a situation in which there are too many nutrients. Algae and plants grow like crazy, then die. Then the decomposers (bacteria and fungi) go to work. In their work of decomposition, oxygen is used up. So in the end, the pond or lake ends up being very low in oxygen, causing animals such as fish to die. A common cause of eutrophication is fertilizer containing phosphorus running off farm fields and into ponds and rivers.

5) Chlorella (which is an autotroph) is eaten by Paramecium (which loves to eat algae) which is eaten by Didinium (which loves to eat Paramecia) which is eaten by Stentor (one of the few protists large enough to eat Didinium).

Activity 3.3:

1) D   2) F   3) C   4) E   5) H   6) A   7) J   8) G   9) K   10) L   11) I   12) B

Activity 3.4:

Technically, there are no "right" answers to this activity. However, here are some suggestions:

yam: Dileptus, Euglena, Loxodes
paper fastener: Chlamydomonas
yellow squash: Dileptus
lollipop: Vorticella
ice cream cone: Stentor

eggplant: Didinium
celery: Spirostomum
slipper: Paramecium
tickets: filamentous algae

necklace: cyanobacteria
green peas: Chlorella
ribbon: Spirpgyra
soccer ball: Volvox

Activity 3.5:

DOWN:

1) mitochondria	5) binary fission	6) spasmoneme	10) Loxodes	11) colonial
12) consumer	13) protist	14) Chlorella	15) micron	16) sessile
17) cytoplasm	20) posterior	22) Paramecium	26) anterior	

ACROSS:

2) trichocyst	3) Golgi body	4) proboscis	7) heterotroph	8) micronucleus
9) pellicle	12) carnivore	18) Halteria	19) Euplotes	21) conjugation
22) prokarayotic	23) Stentor	24) autotroph	25) flagella	27) chloroplast
28) zygote	29) contractile	30) ribosomes		

Chapter 4

Activity 4.2:

GREEK:
dinos: (whirling) dinoflagellate
deinos: (terrifying) dinosaur
saur: (lizard) dinosaur
pyro: (fire) pyrenoid (The reason for this name is unknown.)

LATIN:
luca: (light) Noctiluca ("luca" is similar to "lux")
nocti: (night) Noctiluca
pustula: (blister) pusules (We had to give poor Zeus another word!)

Activity 4.3:

1) all three
2) diatoms
3) all three
4) diatoms
5) dinoflagellates
6) desmids
7) diatoms
8) diatoms, dinoflagellates
9) all three
10) diatoms, dinoflagellates
11) all three
12) all three
13) diatoms
14) diatoms, dinoflagellates
15) all three
16) desmids
17) dinoflagellates
18) dinoflagellates

Chapter 5

Activity 5.1:

GREEK:
amoibe: (change) ameba
gaster/gastro (stomach) gastrotrich
helios (sun) heliozoan
ostrakon (shell) ostracod
pseudes (false) pseudopod
podos (foot) pseudopod
pherein (to bear) rotifer and foraminiferan

LATIN:
roto/roti (wheel) rotifer
foramen (hole) foraminiferan

Activity 5.2:

1) a   2) c   3) d   4) d   5) c   6) a   7) b   8) b   9) d   10) c   11) d   12) c   13) b   14) b   15) e

Activity 5.3:

1) Heliozoan because "helio" is Greek for "sun."
2) Ostracod because they are in class Crustacea, which also contains crabs and lobsters.
3) Shelled ameba because their shells look like helmets and protect the closest thing they have to a head-- a nucleus.
4) Gastrotrich because "gastro" is Greek for "stomach" or "gut."
5) Slime mold because lots of them swarm together to form a large "slug."
6) Foraminiferan because lots of them are found in the limestone that was used to build the pyramids.
7) Rotifer because "roti" is Latin for "wheel."
8) Ameba because they are amorphous and do look like blobs.
9) Radiolarians because billions of them are found in the ooze at the bottom of the ocean.

Activity 5.4:

1) hermaphrodite   2) parthenogenesis   3) rotifer   4) endosymbiont   5) spores
6) trick   7) pseudopods   8) crustaceans   9) silicon
Riddle #1: A test   Riddle #2: Father's Day   Riddle #3: Cytoskeletons   Riddle #4: Their houses are made of glass.

Epilogue

Activity E.1:

ACROSS: 1) flagella   4) cyst   5) Trypanosoma   7) diarrhea   10) parasite   11) saliva   13) mosquito
16) Apicomplexa   19) Tsetse   22) endosymbiont   23) boil
DOWN: 1) falciparum   2) Lyme   4) chronic   6) trepanning   9) intestine   12) coinfection   14) malaria
15) tissue   17) stool   18) gametes   20) herbs   21) tick

Activity E.2:

1) N   2) W   3) V   4) P   5) J   6) O   7) F   8) T   9) S   10) L   11) Z   12) G
13) C   14) I   15) A   16) M   17) E   18) X   19) Q   20) B   21) K   22) D   23) H   24) Y
25) R   26) U

CPSIA information can be obtained
at www.ICGtesting.com
Printed in the USA
BVHW010247240121
598563BV00005B/43

9 780988 780866